Atomic-Scale Electronics Beyond CMOS

Mircea Dragoman · Daniela Dragoman

Atomic-Scale Electronics Beyond CMOS

Springer

Mircea Dragoman
National Research and Development
Institute in Microtechnology
Voluntari, Romania

Daniela Dragoman
Faculty of Physics
University of Bucharest
Magurele, Romania

ISBN 978-3-030-60562-9 ISBN 978-3-030-60563-6 (eBook)
https://doi.org/10.1007/978-3-030-60563-6

This Springer imprint is published by the registered company Springer Nature Switzerland AG
The registered company address is: Gewerbestrasse 11, 6330 Cham, Switzerland

Preface

The fact that the number of transistors on very large-scale integrated (VLSI) circuits, including microprocessors, is doubling every two years, a prediction known as Moore's law is already a history. This prediction was valid for almost fifty years, but its predicted rhythmicity was lost when the dimensions of the transistors started to be lower than 14 nm. The reduction of transistor dimensions below 5 nm will be very difficult to handle, especially due to the high economic costs to produce equipments to produce VLSI circuits containing each billions of transistors at the level of mass production. Thus, increasing computing performances based on increasing the number of transistors in a chip will be no longer possible as soon as transistors reach the atomic scale.

These simple facts inspired a large number of papers and predictions about the "catastrophes" accompanying the end of Moore's law. There are concerns that many industries based on computation will suffer, since the transistors have reached the atomic scale.

What will follow when Moore's law is ended? Are the atomic-scale devices a barrier for further development of electronics? These basic questions are the subject of this book, and the answer is developed along the entire book. It shows that electronics at atomic scale is being developed and it will produce a revolution in many areas, such as computing, communications and many industries based on them. Electronics at atomic scale is at the beginning; therefore, each chapter shows the state of the art, but also what are the problems to be solved in the years to come.

Atomic-scale electronics is based on many technologies, materials and concepts developed in nanoelectronics during the Moore epoch. Moreover, silicon is a very promising material for many atomic-scale devices. So, the experience of Si technologies and devices will be further exploited.

Of course, electronics at atomic scale will have to change many new materials and concepts. New materials and principles of devices based on quantum mechanics, termed as quantum devices, are the core of atomic-scale electronics. As an example, quantum processors for quantum computation have reached a certain maturity and show spectacular results at very low temperatures.

Chapter 1 is opening the book with materials at atomic scale. Graphene and many other 2D materials are presented, focusing on their growth and applications. Single dopants and dangling bonds in semiconductors are treated also in this chapter since their applications are important to atomic-scale electronics. Chapter 2 is dedicated to electronics devices at atomic scale, i.e., devices based on a single or few dopants, gap atomic switches, atomic switches and memristors based on them. The chapter ends with the description of nanogaps and nanoribbons and the devices based on them.

Chapter 3 is the largest of the entire book and is dedicated to nanoelectronic devices developed during the search of alternative developments of Moore's law. Here, we present ballistic devices, negative capacitance FETs, hyper-FETs, tunneling devices, phase change devices, quantum dots, memristors and memories based on them. Many of them have in common the fact that quantum mechanics is at the ground of their functionalities.

The next two chapters are dedicated to quantum and neuromorphic computation, focusing on devices that make such advanced computation possible. In Chap. 4, the physical basis of quantum computing is first explained. Then, we provide few practical examples, linked to electronics at atomic scale, of quantum computing based on superconducting Josephson junctions and quantum dots. The knowledge which is gained in the previous chapters, such as tunneling effects, quantum dots, SETs and FETs, is used to understand the essence of a quantum processor, which is the follower of microprocessors used in digital computers. Chapter 5 is focused on neuromorphic computation. Such advanced computation is based on the analogies of human brain functioning. In this type of computation, artificial neurons and synapses are the key electronic devices. Because memristors are artificial synapses, this device is explained in the context of neuromorphic computation, while in Chap. 3 its role in memories was detailed. This type of computing is dedicated to artificial learning and artificial organs such as retinas and bionic eyes.

The book ends with a short chapter describing the perspectives of electronics at atomic scale.

The main goal of the book is to show that electronic devices having atomic-scale dimensions are not at the end of the development of nanoelectronics, computing or communications. On the contrary, the electronics at atomic scales opens new ways in the development of many areas of applied sciences and industries in the years to come.

Bucharest, Romania

Mircea Dragoman
Daniela Dragoman

Contents

Chapter 1
Materials at Atomic Scale

Abstract This chapter presents the materials that are used in electronics at atomic scale. It starts with atomically thin materials, which are monoatomic layers, also known as atomically thick materials or 2D materials. These include the most widespread 2D materials: graphene and transition metals chalcogenides (TMDs). The chapter continues then with single dopants in semiconductors, which are used in atomic electronics, and dangling bonds used in logic gates.

1.1 Atomically Thin Materials

1.1.1 2D Materials Growth and Electrical Contacts

There are more than 200 atomically-thin materials (Miró et al. 2014), divided in several families:

X-enes, which are monoatomic layers formed from a single type of atoms arranged in a honeycomb lattice. In particular, the monoatomic layer formed from carbon atoms is termed as graphene—the most famous 2D material, which is intensively studied for its applications in electronics, photonics, or biology (Dragoman and Dragoman 2017). On the contrary, if the monoatomic layers are formed from Si or Ge atoms, we obtain silicene and germanene, which are much less studied since they are very difficult to grow and are unstable in air. When the monoatomic layer is formed from phosphorus atoms, the 2D material is termed as phosphorene; phosphorene originates from/is a 2D version of black phosphorous, as graphene originates from graphite. Although it is unstable in air, 2D black phosphorous (BP) has many applications in electronics and energy applications. Hexagonal boron nitride (h-BN) is the monoatomic thin counterpart of BN, presenting in-plane sp^2 covalent bonding. Hexagonal boron nitride has a paramount importance in heterostructures involving graphene, since its hexagonal lattice is similar to that of graphene. A recent review about monoatomic layers can be found in D'Angelo and Matsuda (2019). Graphene, germanene and silicene are semimetals, i.e., they

M. Dragoman and D. Dragoman, *Atomic-Scale Electronics Beyond CMOS*,
https://doi.org/10.1007/978-3-030-60563-6_1

are semiconductors with a zero bandgap, while phosphorene is a semiconductor with a bandgap of 0.3 eV and h-BN is an insulator with a bandgap of 6 eV.

TMDs (transition metal dichalcogenides), which have the generic form MX_2, are 2D materials constituted from a transition metal (M) monolayer such as Mo, W, Ti, Hf, Fe, Co which is covalently bonded with two chalcogenide monolayers (X_2) such as S, Se, Te. The most studied materials from this family due to their promising applications in electronics and photonics are MoS_2 and WS_2, which are atomically thin semiconductors.

SMCs (semimetal chalcogenides), which are semiconductors with the generic formula M_2X_2 where M denotes a semimetal, such as In or Ga, and X stands for a chalcogenide, such as Se or S.

MX-enes have also a hexagonal lattice and are described by the formula $M_{n+1}AX_n$, $n = 1, 2, 3$, where M is a transition metal, A is an element from group 13 or 14, and X denotes carbon or hydrogen. Their applications are growing, being for example excellent for electromagnetic shielding in microwaves.

All these materials are van der Waals (vdW) materials (see Fig. 1.1), i.e., layered materials with strongly anisotropic bonds. More precisely, in the plane of the layer there are strong covalent bonds (with binding energy of the order of eV/atom), while perpendicular to the layer plane there are weak bonds (with corresponding binding energy three order of magnitude weaker, of meV/atom). An example is graphite with 7.3 eV/atom binding energy in the layer plane and only 60 meV/atom perpendicular to this plane, the interlayer distance being of 0.34 nm (Duong et al. 2017). Thus,

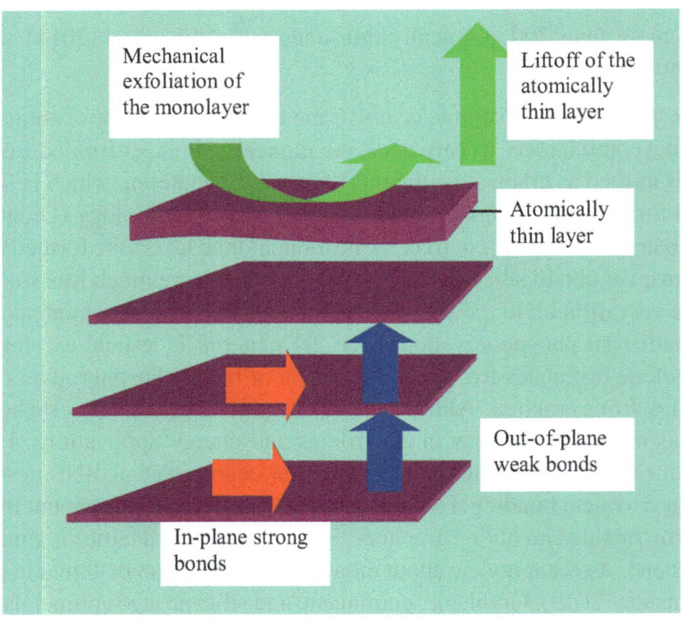

Fig. 1.1 Van der Waals materials

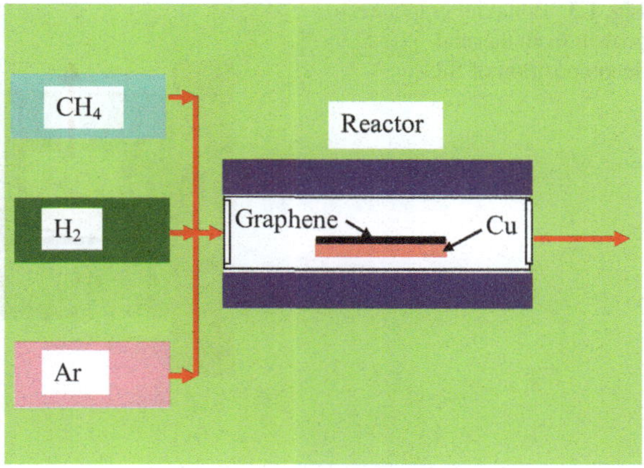

Fig. 1.2 Graphene growth by CVD method

atomic planes, termed as monolayers, or in case of graphite termed as graphene, can be quite easily mechanically exfoliated, for example with a Scotch type. This is still the most popular method even today to obtain 2D materials, but without relevance for nanoelectronics, where a large number of devices are required on every inch.

There are different methods to grow 2D materials, but we will focus here only to growth methods able to produce atomically thin materials at the wafer level, which are the only methods of interest for producing advanced electronic devices. For graphene, a widespread growth method at the wafer level is chemical vapor deposition (CVD) (Whitener and Sheehan 2014). In this case, a carbon precursor in the gaseous phase, such as methane or ethylene, is introduced in a reactor chamber at high temperature (1000 °C) and the graphene is formed on the surface of a catalyst located inside it (for example, Cu in Fig. 1.2), and then transferred on a substrate. The utilization of isotopes of methane such $^{12}CH_4$ and $^{13}CH_4$ allows tracking of carbon during graphene growth (Frank et al. 2014). However, grain boundaries appear during CVD growth, such that large islands of graphene monolayers are surrounded by regions with mixtures of graphene types (bilayers, trilayers and multilayers of graphene are found).

Another method to grow graphene at wafer scale is epitaxial growth by thermal decomposition of SiC at high temperatures. In this case, in ultrahigh vacuum and at temperatures above 800 °C, silicon sublimates from the SiC wafer and a high quality epitaxial graphene layer is obtained (de Heer et al. 2011). The hexagonal SiC wafer in high vacuum is surrounded by a graphite enclosure which is heated, as schematically represented in Fig. 1.3.

The graphene monolayer can be grown on either the Si or the C face of SiC, graphene having different properties in these two instances. There are two main limitations of graphene growth on SiC:

(i) the difficulty of a precise control of sublimation in the range 900–1300 °C, and

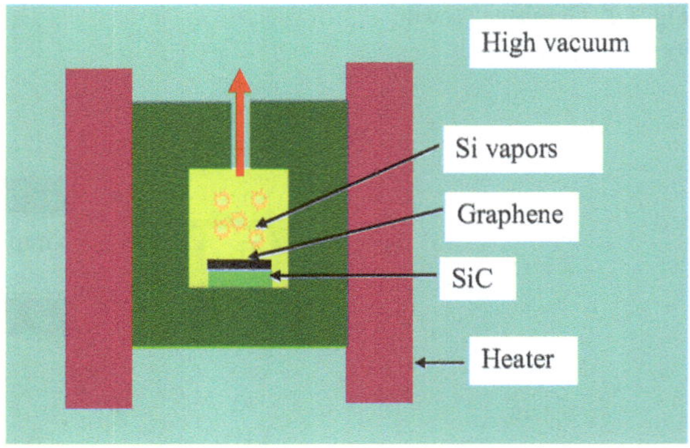

Fig. 1.3 Epitaxial graphene growth from thermal decomposition of SiC

(ii) the limitation of thermal decomposition of Si to the surface 3C-SiC(111), but there are new methods to improve the epitaxial graphene growth at the wafer scale based on Si atoms decomposition at SiC surface (Mishra et al. 2016).

For nanoelectronic applications, both methods above must be accompanied by transfer of graphene on a certain substrate, process which adds additional defects in graphene. However, there are modern methods to grow graphene without transfer methods, or transfer-free graphene growth, starting from a metal or dielectric surface (Kaur et al. 2019). An example of transfer-free graphene growth at the wafer level is displayed in Fig. 1.4.

The above-mentioned method is important because during transport additional defects appear, while water, solvents, or resists can be trapped between the graphene and the dielectric, limiting its performances, especially the electron mobility. The problem of transfer-free CVD graphene growth is that the substrate could be damaged during growth at high temperatures, of nearly 1000 °C. Therefore, the reduction of the temperature in the process is a big demand, realized only recently (Fujita et al. 2017) using diluted methane and molten Ga as catalyst, allowing the reduction of the temperature down to 50 °C on the sapphire substrate and down to 100 °C on polycarbonate. The growth process is possible via carbon adhesion to previously produced graphene islands. Gallium, which is a liquid metal at room temperature, is used as catalyst in CVD growth due to several advantages: (i) graphene is not soluble in Ga, (ii) the lattice mismatch is improved, and (iii) Ga improves graphene growth speed, allowing thus the reduction of the temperature. Figure 1.5 depicts schematically a typical growth process.

The CVD method can be used to grow other various 2D materials at the wafer scale, such as h-BN or TMDs. Any 2D material growth process implies several stages before forming a continuous film at the wafer scale, which are displayed in Fig. 1.6 (Jung et al. 2019).

The h-BN is grown at large scales, even at wafer scale, and with controllable orientation using low-pressure CVD (LPCVD) (Song et al. 2015). Ammonia borane (BH3–NH3) is used as a precursor in this case. The h-BN monolayers have a paramount importance in 2D electronics since their lattice structure is similar to graphene,

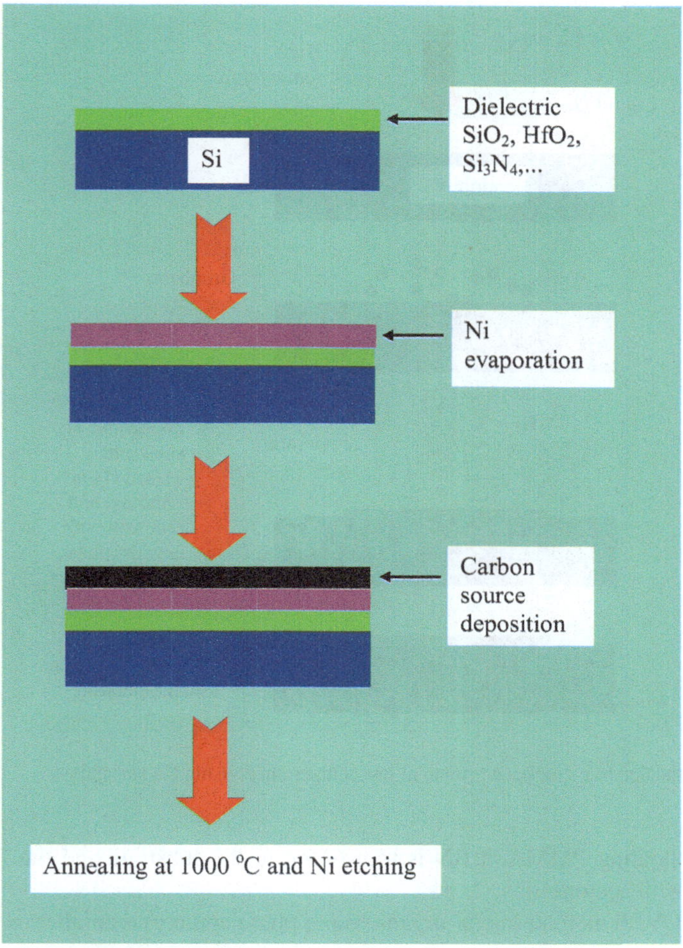

Fig. 1.4 Principle of CVD transfer-free graphene deposition

although the material is a dielectric (the bandgap of h-BN is 5.9 eV). In addition, the h-BN surface is flat, free of dangling bonds, and thus eventual charges trapped at the interface of h-BN with graphene or other 2D materials are reduced, so that the mobility can increase an order of magnitude. The CVD growth of h-BN is depicted in Fig. 1.7. The ammonia borane precursor is introduced in the CVD tube via a special quartz cell, where it first sublimates by heating in a heating belt, before the resulting species enter the reaction tube. The h-BN is grown on Cu foil and then transferred on the desired substrate, for instance Si/SiO_2, using a wet-etching technique.

The transfer on the desired substrate is done in a similar manner as in the case of graphene grown by CVD, i.e., the h-BN layer is covered with PMMA and immersed in an etchant (68% H_3PO_4, 3% HNO_3 and 5% CH_3COOH in water) (Jang et al. 2016).

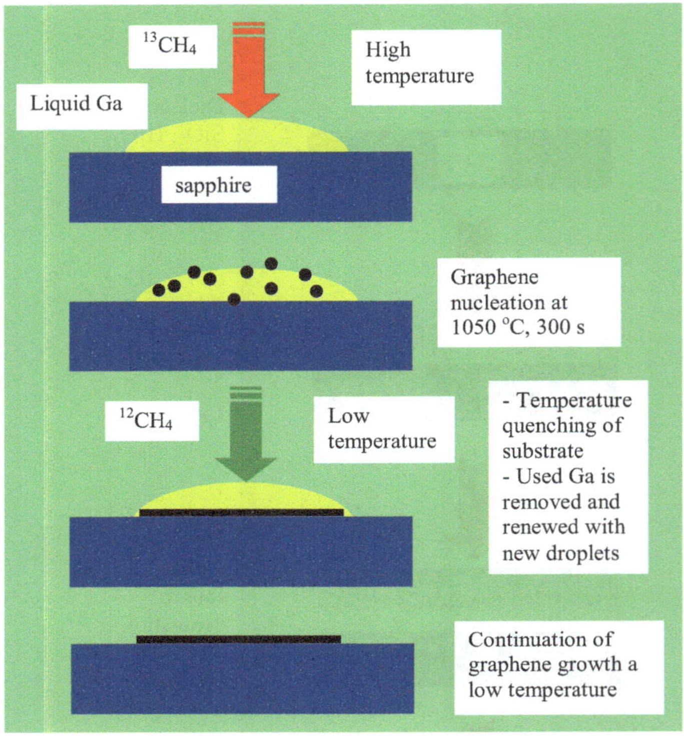

Fig. 1.5 Transfer-free graphene growth at low-temperatures using Ga as catalyst

Then the floating PMMA–h-BN is transferred to the substrate and the PMMA is removed with acetone.

The MOVPE method (metal-organic vapor phase epitaxy) is another method able to grow many 2D materials at the wafer scale. In fact, MOVPE is a well-known method to grow crystalline or polycrystalline thin films, and even heterostructures, the majority of semiconductors and semiconductor heterostructures being grown by this method, and is now extended with success for 2D materials growth.

The growth of 2D materials using MOVPE relies on chemical reactions of gases and metalorganic (MO) precursors over a substrate, at high temperatures and rather moderate pressures. For example, in the case of h-BN, the precursors for B and N, respectively, are Trethylboron (TEB) and NH_3 (Sundaram et al. 2019), while the metal-organic precursors for MoS_2 growth are molybdenum hexacarbonyl (MHC), $Mo(CO)_6$, and hydrogen sulfide, H_2S, and the substrate is *c*-plane sapphire (Kwak et al. 2019). The MOVPE growth of MoS_2 is represented in Fig. 1.8.

The evolution of the deposition/growth process in this case is verified via Raman spectrum, which has a special imprint for MoS_2. More precisely, there are two bands, located at 400–410 cm^{-1} and 380–390 cm^{-1}, showing two peaks related to the A_{1g} and E_{2g} modes, respectively. The frequency difference between the peaks of the A_{1g}

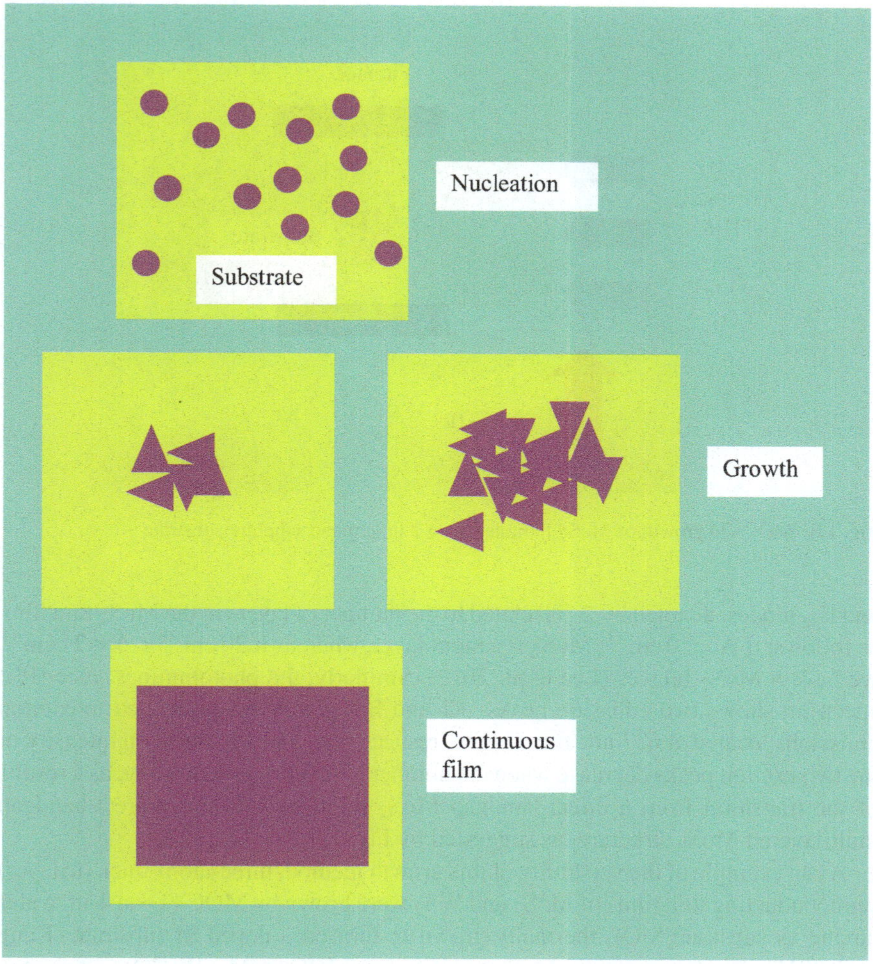

Fig. 1.6 Stages of a 2D material growth using CVD

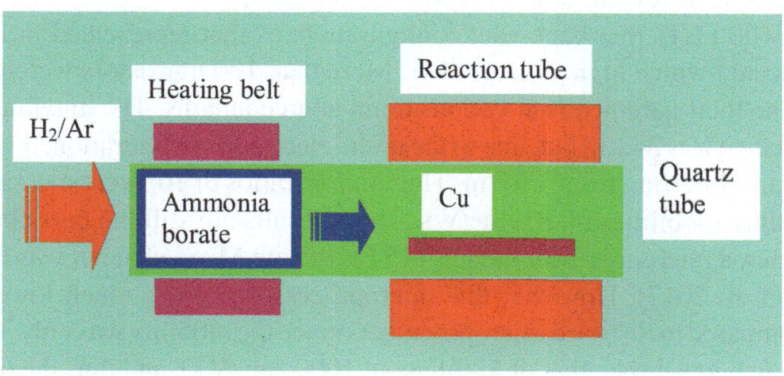

Fig. 1.7 CVD growth of h-BN

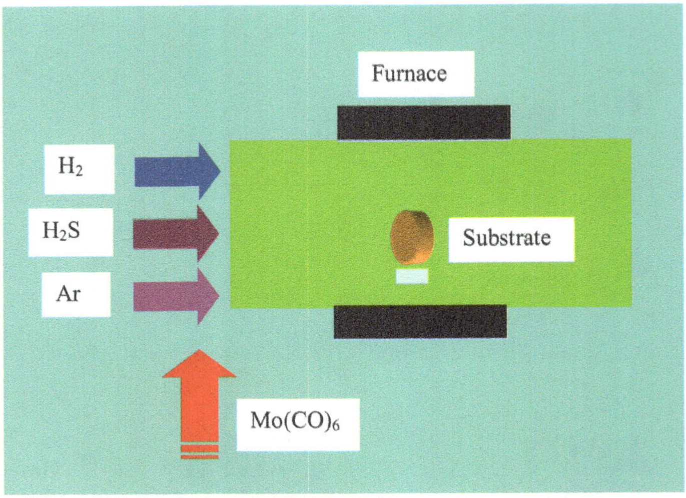

Fig. 1.8 MOCVD growth of MoS_2 monolayer on 2 in. *c*-plane sapphire substrate

and E_{2g} modes, denoted by Δ, is related to the number of layers in the MoS_2 thin film, as follows: if $\Delta < 20\ cm^{-1}$, MoS_2 is a monolayer, whereas if $20\ cm^{-1} < \Delta < 21\ cm^{-1}$ we have a MoS_2 bilayer (Lee et al. 2012). Similarly, the photoluminescence (PL) spectrum shows two emission peaks, A1 and B1, which are associated to exciton emissions located at 627 nm and 677 nm, respectively. The PL emission intensity of the two exciton peaks decrease when the number of layers are increasing, as a results of the transition from a direct-bandgap MoS_2 monolayer to an indirect-bandgap multilayered MoS_2 structure, as suggested by Fig. 1.9.

As an example of the versatility of this growth method, three-atom-thick (trilayer) semiconducting thin films of MoS_2 and WS_2 were grown by MOCVD on 4-in. wafer having as substrate SiO_2, thermally grown in turn on a doped Si substrate (Kang et al. 2015). The precursors for MoS_2 and WS_2 are: MHC for Mo, $W(CO)_6$—tungsten hexacarbonyl (THC) for W, and $(C_2H_5)_2S$ for S. The growth temperature was 550 °C and the pressure was maintained at 7 Torr during 26 h.

Thus, 8100 FETs in a back-gate configuration, as that represented in Fig. 1.10, were fabricated with a high yield of 99%. More than 100 transistors were measured to determine their parameters at various drain-source lengths. The maximum drain-source distance was 34 μm but, due to the multi-electrode configuration, this distance could be varied starting from 1.6 μm. Thus, *on/off* ratios of 10^6 were measured, with corresponding mobilities of 30 cm^2/V s. These results are still the best obtained up to now. Later, a microprocessor having 115 FETs with MoS_2 channel was fabricated (Wachter et al. 2017). However, this microprocessor showed much lower performances compared to Si-based microprocessors since the mobility was only 3 cm^2/V s, which is much lower than that of Si (of about 1000 cm^2/V s), and thus the transistors have a maximum frequency of only 20 kHz.

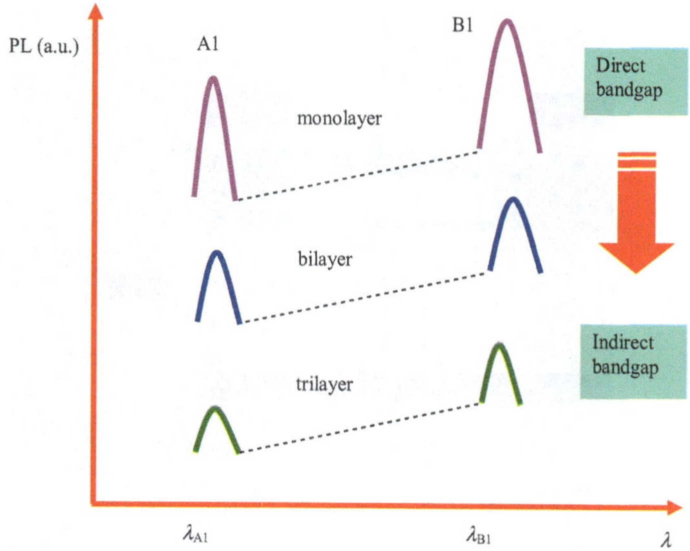

Fig. 1.9 PL imprint of MoS_2 multilayer structures

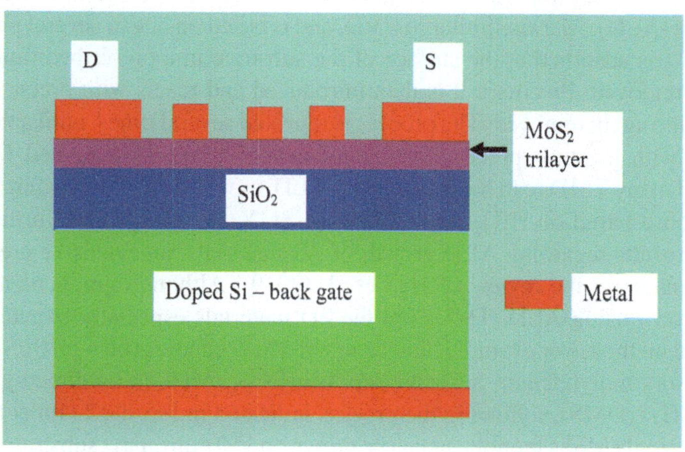

Fig. 1.10 Multi-electrode MoS_2 trilayer FET

Only very recently it was found that the mobility of MoS_2 can reach the value of 820 cm^2/V s (Liu et al. 2019), value close to that of Si, but the substrate is crested, i.e., the MoS_2 multilayer is deposited on a nanostructured substrate, with the surface (SiN_x or HfO_2) displaying corrugations with a r.m.s roughness of 2 nm. This important result is explained by the fact that the mechanical strain is increasing the mobilities of charge carriers in TMDs. A crested back-gate TMD FET is shown in Fig. 1.11.

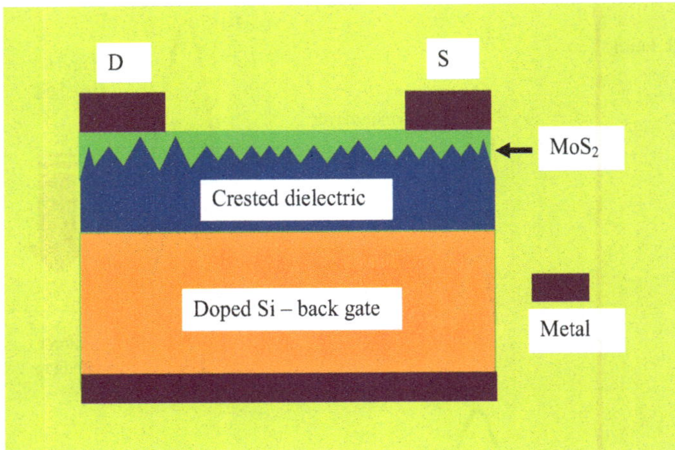

Fig. 1.11 Crested TMD FET transistor

Because the mobility is the key issue of the development of electronics at atomic scale, we resume the discussion of growth methods of 2D materials. One deposition method with a high potential to obtain 2D materials at the wafer level is atomic layer deposition (ALD). ALD is similar to CVD, and is based on one or several precursors. Precursor 1 is adsorbed at the surface of the substrate in a prescribed time and the excess is removed. Precursor 2 is then introduced and reacts with precursor 1, the excess being again removed. The process is repeated several times, until a thin film is deposited with angstrom precision, i.e., until an atomic layer is deposited. Graphene, h-BN, and many TMDs can be deposited by ALD, as well as few-layers ferroelectrics such as those based on HfO_2 doped with Zr, Si, Y, etc. We present further briefly the main results regarding ALD growth of 2D materials, an extensive presentation being available in the review of Hao et al. (2019). Although graphene and h-BN can be grown using ALD, TMDs are the 2D materials especially targeted by this deposition method. For example, few layers of MoS_2 can be grown on Si/SiO_2 using as precursors molybdenum hexacarbonyl $Mo(CO)_6$ (MHC) and dimethyldisulfide (CH_3SSCH_3) (DMS), following the process indicated in Fig. 1.12 (Jin et al. 2014), whereas a single MoS_2 monolayer can be grown on 150 mm quartz substrate, showing the power of the ALD deposition method (Valdivia et al. 2016).

There are many more 2D materials beyond graphene and TMDs, which are now grown via epitaxial methods (CVD, MBE PVD) (Geng and Yang 2018). Among them, and apart from boron-based materials such as h-BN, discussed above, there are many other 2D boron-based materials, such as borophene. Borophene is a 2D sheet of boron atoms showing interesting physical properties, but rather difficult to be grown. A straightforward method to grow borophene using CVD on a copper foil has produced sheets of this material with dimensions of few centimeters (Tai et al. 2015). We point out that graphene is grown on CVD also on a copper foil. The CVD process is described in Fig. 1.13. The resulting 2D material, with a thickness of

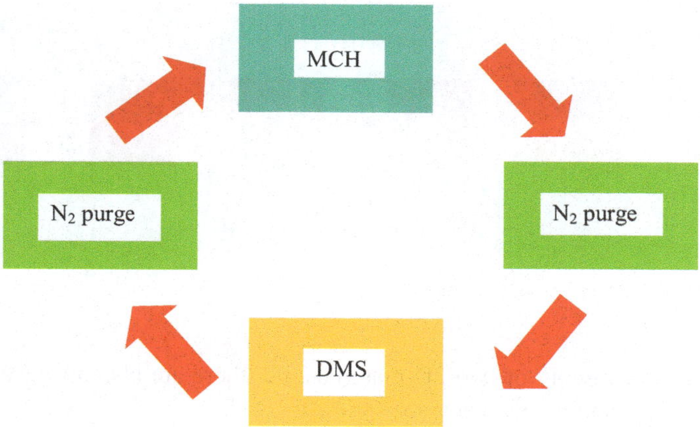

Fig. 1.12 The ALD process for MoS_2 deposition

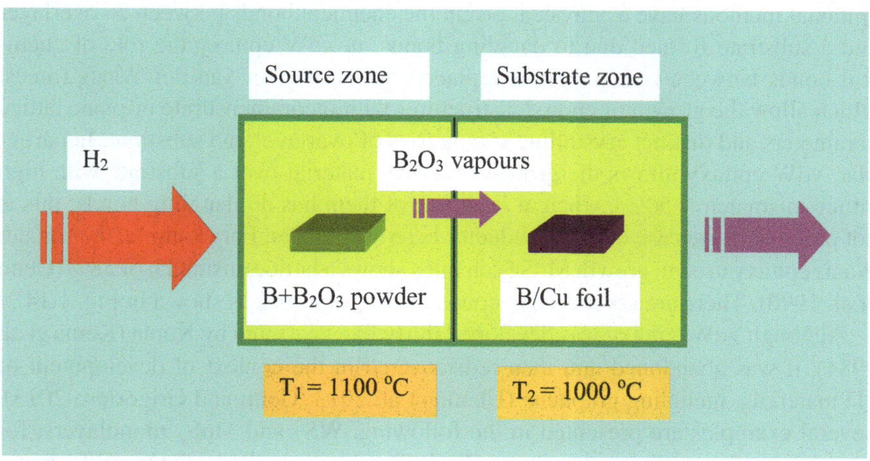

Fig. 1.13 CVD growth of borophene on a Cu foil

0.8 nm, is in fact the polymorph γ-B_{28}. Its structural analysis reveals a complicated structure formed of orthorhombic cells that contain icosahedral B_{12} units and B_2 dumbbells.

Two-dimensional analogues of graphene obtained from other group-4 materials than C, namely Si and Ge, are silicene and, respectively, germanene. They are very difficult to be grown and are unstable in air. The same situation applies to phosphorene, which is the 2D version of phosphorous.

Two-dimensional materials are able to form heterostrostructures, similar to the case of bulk semiconductors, and in this way are able to produce band-engineered structures, which is a key concept in modern nanoelectronics. Many studies are based

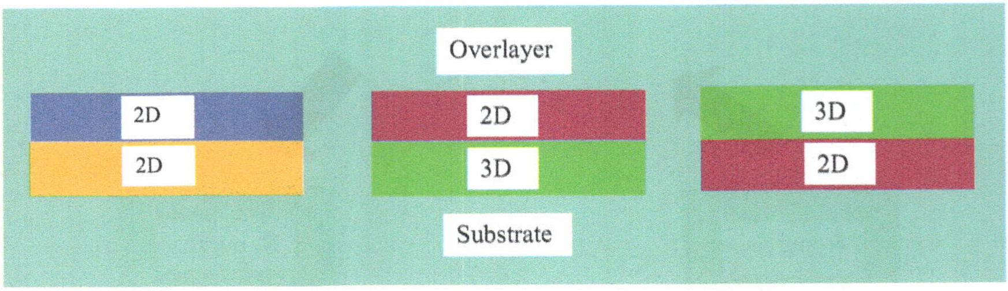

Fig. 1.14 The vdW epitaxy configurations

on mechanical assembly of two 2D materials, but there are also advanced growth techniques for obtaining 2D heterostructures.

One of these is van der Waals (vdW) epitaxy (Utama et al. 2013). The vdW epitaxy is based on vdW forces between an overlayer and a substrate. The vdW epitaxy is rather different from the epitaxial methods described above. More precisely, whether epitaxial methods have as physical origin the chemical bonds between an overlayer and a substrate formed due to dangling bonds, in vdW epitaxy the role of chemical bonds between monolayers is replaced by the weaker van der Waals forces, which allow the growth of epitaxial structures with incommensurate in-plane lattice parameters and distinct crystalline symmetries of overlayer and substrate. In particular, vdW epitaxy allows the growth of a new material over a substrate with high lattice mismatch (>50%), when at least one of them has no dangling bonds; this is not possible in the case of semiconductor heterostructures. For example, the van der Waals epitaxy used to growth $MoSe_2$ on mica shows a lattice mismatch of 58% (Ueno et al. 1990). There are several vdW epitaxy configurations, as shown in Fig. 1.14.

Although vdW epitaxy was developed thirty five years ago by Koma (Koma et al. 1984), it was abandoned and then rediscovered in the context of development of 2D materials, including graphene (Utama et al. 2013; Geim and Grigorieva 2013). Several examples are presented in the following. WS_2 and MoS_2 monolayers, for instance, can be grown separately by CVD (Tongay et al. 2014), the top WS_2 material being deposited afterwards over the MoS_2 via the stamping method. This is a vertical vdW heterostructure, having a Si/SiO_2 substrate. An interesting aspect is that the Raman and PL spectra of the heterostructure are composed of the Raman spectrum and PL spectrum of each monolayer, which are contributing separately to the total spectra. This is an unusual effect, which is not encountered in semiconductor heterostructures, and originates from the weak interaction between different layers. These weak vdW forces manifest also in large interlayer separations that can be reduced by annealing.

The vdW heterostructures can be grown by CVD using a method that employs a couple of furnaces (Wang et al. 2015). For example, h-BN is grown by CVD on Cu foil and transferred on a Si/SiO_2 chip. Then, the chip is placed into a second furnace, where MoS_2 is CVD grown over h-BN. However, the precursors of CVD for MoS_2 are located in another furnace. Different temperatures are used for MoO_3, S and substrate (Fig. 1.15).

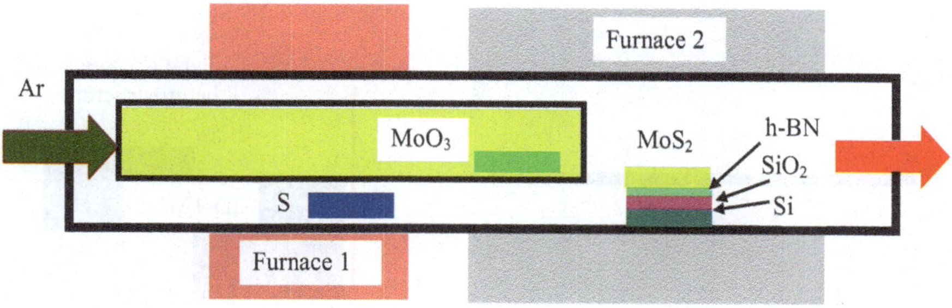

Fig. 1.15 vdW growth of the MoS_2/h-BN heterostructure

Fig. 1.16 PVD growth of the WS_2/MoS_2 heterostructure

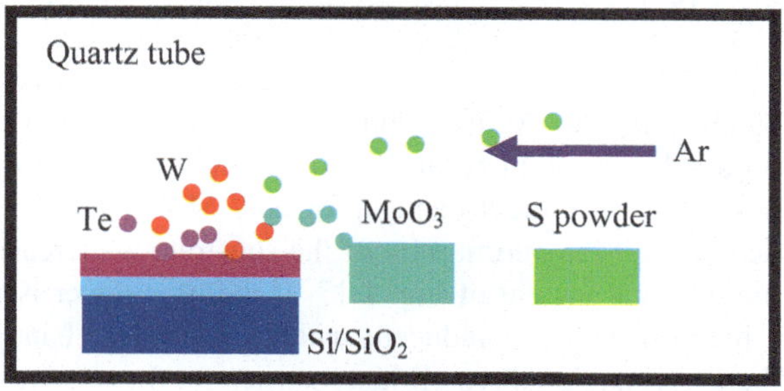

It is interesting to note that WS_2/MoS_2 vertical and in-plane heterostructures could be grown in a single step via the physical vapor deposition method (PVD) by very tight control of temperature, as shown schematically in Fig. 1.16 (Gong et al. 2014). Now, the MoO_3 powder is located near the Si/SiO_2 chip, on the surface of which W and Te are dispersed. In contrast, the S powder is located in a zone with decreased temperature and far from the chip, and S is brought to the chip via the Ar gas. Due to their different nucleation rates, WS_2 and MoS_2 are sequentially grown.

There are not only vertical heterostructures, but also lateral heterostructures when the 2D materials have a common boundary in the same plane. This latter type of heterostructures can be produced even when we use a single 2D material and two dielectric substrates, one with higher and the other with lower dielectric permittivity. In 2D materials the screening is reduced due to strong Coulomb interactions, and thus the screening and the bandgap could be engineered by the different substrates with different permittivity values (see Fig. 1.17). In Fig. 1.17, E_c and E_v denote the bottom and top, respectively, of the conduction and valence bands across the resulting type II staggered heterostructure.

In general, in 2D materials the bandgap is no longer a material parameter, but is drastically changed by the surroundings, such as substrate, the top dielectric above the 2D material, or by both situations. The effect is referred to as dielectric-induced bandgap renormalization (Cho and Berkelbach 2018) and this effect is opening a novel way of bandgap engineering, which is not encountered in semiconductors. For example, in the case of the MoS_2 monolayer, when suspended in air the direct

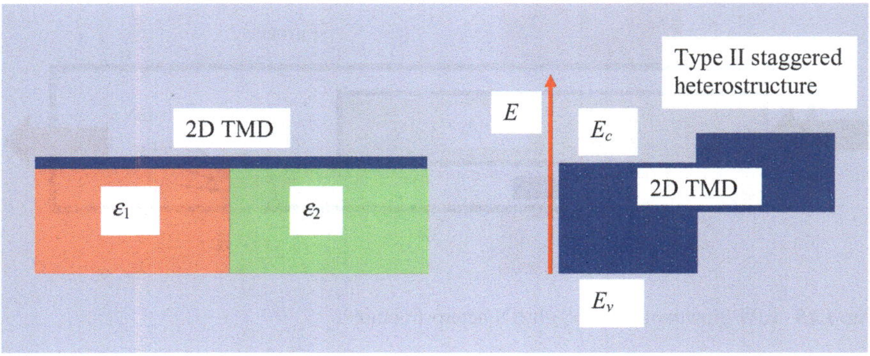

Fig. 1.17 Induced lateral heterostructure via screening

bandgap is 2.62 eV (the relative substrate permittivity ε_r is 1), but if MoS_2 is grown on a substrate with ε_r of 20 (for example, HfO_2) the bandgap becomes 1.95 eV; note that the indirect bandgap of bulk MoS_2 is 1.98 eV. In principle, the bandgap decreases if the permittivity of the substrate is increasing. Now, we can understand the physical insight of Fig. 1.17. If the monolayer is deposited on two dielectrics, a heterostructure is induced, with two different bandgaps with an offset between them—a staggered type II heterostructure, for relative permittivities $\varepsilon_1 > \varepsilon_2$.

Very recently, the concepts above were tested via a lateral heterostructure based on two dielectrics (Utama et al. 2019): cytop, having $\varepsilon_r = 2$, and h-BN, with $\varepsilon_r = 6$, which induce a bandgap offset of 90 meV (see Fig. 1.18). Here, the relation between the relative permittivities of dielectrics: $\varepsilon_1 < \varepsilon_2$ for cytop and h-BN, respectively, tells us that we are dealing with a type I heterostructure (see Fig. 1.19).

Fig. 1.18 VdW lateral heterostructure based on two dielectrics

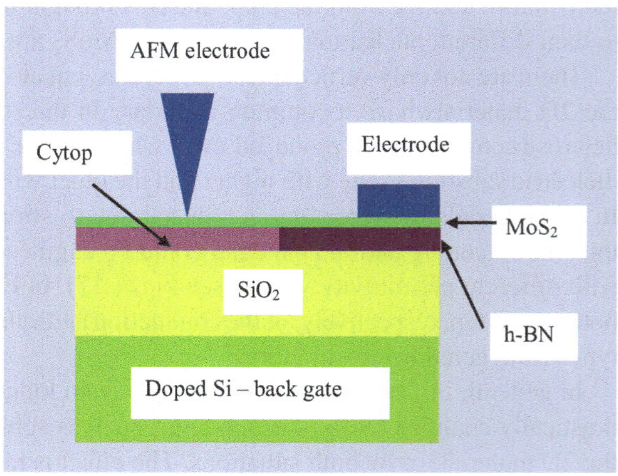

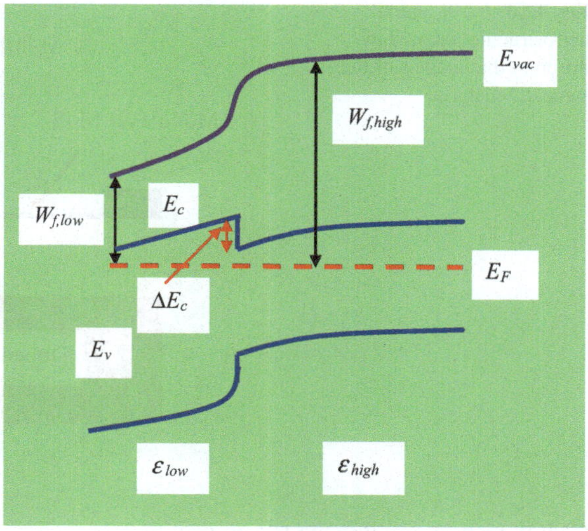

Fig. 1.19 Band diagram of the vdW lateral heterostructure based on two dielectrics (after Utama et al. 2019)

The band offset is calculated using the formula:

$$\Delta E_C = -\Delta W_f - k_B T \ln\left(\frac{\exp(\pi \hbar^2 n_{\text{low}}/mk_B T) - 1}{\exp(\pi \hbar^2 n_{\text{high}}/mk_B T) - 1}\right) \tag{1.1}$$

where $\Delta W_f = W_{f,\text{low}} - W_{f,\text{high}}$ is the workfunction difference between the two dielectrics, $n_{\text{low,high}}$ denote the carrier densities in the two regions, and m is the effective mass of charge carriers. Such heterostructures could have important applications in the future since they work like a bandgap transistor.

Another heterostructure of type I, similar to that presented above, can be produced by varying the thickness of TMDs, such as MoS_2 (Tosun et al. 2015). Such a heterostructure is presented in Fig. 1.20. It is known that TMD monolayers have a direct bandgap, while beyond few layers the bandgap is indirect and is decreasing with the number of layers. Further details regarding vdW growth and its applications are found in the review (Duong et al. 2017).

A special issue less treated in the literature regards the electrical contacts in 2D materials. Many physical properties of these materials, which will be further described, cannot be completely fructified without a low contact resistance.

In principle, in 2D semiconductor materials, the limit of the contact resistance is (Allain et al. 2015):

$$r_c = h/2e^2/k_{\text{F}} = 0.026\sqrt{n_{2D}} \tag{1.2}$$

where k_{F} is the Fermi level wavenumber and n_{2D} is the charge carrier density. If $n_{2D} = 10^{13}\ \text{cm}^{-2}$, then $r_c = 0.030\ \Omega$ mm, which is the state of the art in Si technology, where $r_c < 0.01\ \Omega$ mm (Schwierz et al. 2015). Only graphene monolayers and bilayers with contact resistances between 0.01–0.2 Ω mm could satisfy the requirements of

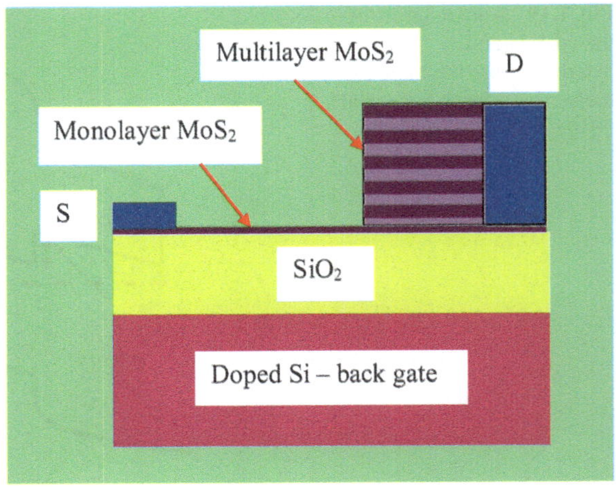

Fig. 1.20 VdW lateral hetrostructure created by different thicknesses of the same 2D material

CMOS technology if using top contact metals such as Ti, Ni, Cr/Au (Schwierz et al. 2015). For comparison, TMDs have contact resistances higher with at least two orders of magnitude compared to Si.

Three large categories of contacts are used for 2D materials: top contacts, lateral contacts and edge contacts (Cheng et al. 2018), as illustrated in Fig. 1.21. The top metal contact is the most widespread type of contact and several approaches have been developed to improve the contact resistance in its case, such as the use of different types of metals, the introduction of an interlayer between the metal and 2D materials, the use of doping etc. The lateral contact is used when graphene acts as electrode and a TMD as a semiconductor. The contact is ohmic, but the contact resistance value is not low. Another type of lateral contact is encountered between the same TMD in two phases, such as 2H MoS_2, which is a semiconductor, and 1H MoS_2, which is metallic, but this phase engineering approach takes place at temperatures higher than 600 °C, which is not compatible with CMOS technology. The edge contacts are very promising, such as Cr edge contact to graphene, for which r_c is of 0.15 Ω mm (Wang et al. 2013), but the method is not widespread to other TMDs.

The contact resistance is given by (Allain et al. 2015):

$$r_C = (\rho_{2D}\rho_C)\coth[l(\rho_{2D}/\rho_C)] \tag{1.3}$$

where ρ_C [Ω m^2] is the resistivity of the metal/2D interface, ρ_{2D} [Ω/square] is the sheet resistivity, and l is the contact length. Another parameter of interest is the transfer length, which is the distance over which carriers are transported below the contact before reaching the contact. It is defined as $L_C = (\rho_C/\rho_{2D})$. If the contact length $l \ll L_C$, we have:

$$r_C = \sqrt{\rho_C\rho_{2D}} \tag{1.4}$$

and the contact resistance is independent of l.

Fig. 1.21 Electrical contacts in 2D materials: (a) top, (b) lateral, and (c) edge

Why in 2D materials we have much higher r_C values than in CMOS devices? In the case of top contacts, there is a so called vdW gap between the metal and the 2D material, since there are no bonds on the surface of 2D materials. In this situation, after the carriers are tunneling through the vdW gap, a Schottky contact develops behind it. Thus, a significant number of carriers are not able to tunnel and subsequent cross over the Schottky barrier, the current is reduced and the contact resistance is increased (see Fig. 1.22).

Therefore, intentionally produced defects can be introduced in the contact region to create the missing bonds and to reduce the contact resistance. In the case of defects creating bonds between a metal and a 2D TMD semiconductor, the vdW gap could vanish and the Schottky barrier is reduced. Then, if the transport of charge carriers at the metal/2D TMD interface can be described as diffusive, the interface is modeled as a resistive circuit, as indicated in Fig. 1.23.

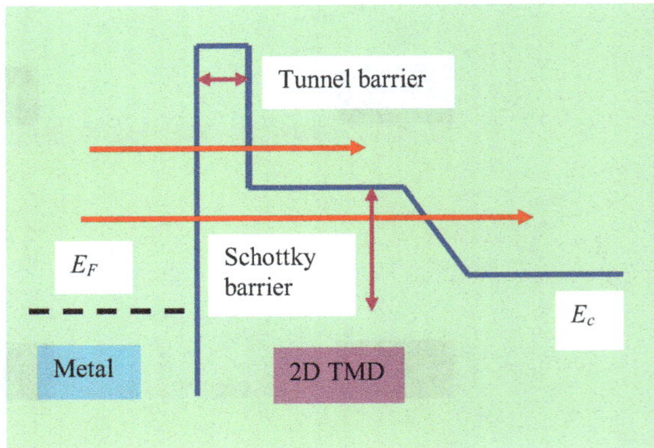

Fig. 1.22 The metal/2D junction

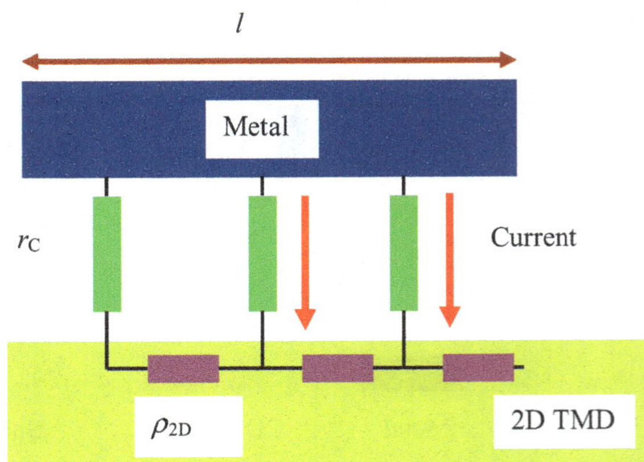

Fig. 1.23 Current injection from metal to a 2D TMD semiconductor

There is another problem related to Schottky contacts, concerning the Fermi level, which is pinned at the semiconductor interface state energy Φ_{IS}. The Schottky barrier height Φ_{SB} is given by (Schulman et al. 2018):

$$\Phi_{SB} = (S\Phi_M - \chi_S) + (1 - S)\Phi_{IS} \tag{1.5}$$

where Φ_M is the workfunction of the metal contact, χ_S is the semiconductor affinity, and S is the Schottky pinning factor given by

$$S = \partial\Phi_{SB}/\partial\Phi_M \tag{1.6}$$

The value $S = 0$ corresponds to a strong pinning effect, while $S = 1$ is the limit of a normal Schottky contact, reaching the Bardeen limit where the Fermi level is located at Φ_{IS}. In the case of MoS_2, $S = 0.1$ as resulting from many measurements on different metals. The Fermi pinning effect reduces the Schottky barrier height. For example, in the case of scandium, the Schottky barrier should be 0.7 eV in the absence of pinning, but it is found to be of only 30 meV. In the case of Pt, the discrepancies are maintained: the Schottky barrier should be 1.4 eV, but it is only 230 meV. Thus, Sc, which has a small workfunction (3.6 eV), is forming a very low Schottky barrier, whereas metals with high workfunctions form higher Schottky barriers. For example, gold and platinum, with workfunctions of 5.5 and 5.6 eV, are forming Schottky barrier heights of 0.2 and 0.23 eV, respectively.

Lower Schottky barriers are in principle and advantage, but in the case of 2D materials this decrease of barrier heights is due to defects at the interface of the 2D semiconductor with the metal, having as consequence that other properties of the 2D material, such as the mobility, are depreciated. Although many efforts have been invested to improve the contact resistance of 2D materials, this issue is still open, and downgrades the performances of 2D devices.

1.1.2 Physical Properties of 2D Materials

After studying the growth of 2D materials and the thorny issues related to metal/2D materials contacts, let us briefly discuss their physical properties. We start with graphene, the most well-known 2D material.

Graphene monolayers can be seen from a physical point of view as native 2D electron gas materials, where the transport of carriers is described by the Dirac equation. The graphene monolayer is a semiconductor with zero bandgap and it is among the few materials where the transport of carriers is not described by the Schrödinger equation (Dragoman and Dragoman 2009a, b).

The dispersion relation $E(k)$ in graphene monolayers is linear at low energies (see Fig. 1.24) and in high-quality graphene, and is given by:

Fig. 1.24 The graphene monolayer dispersion relation

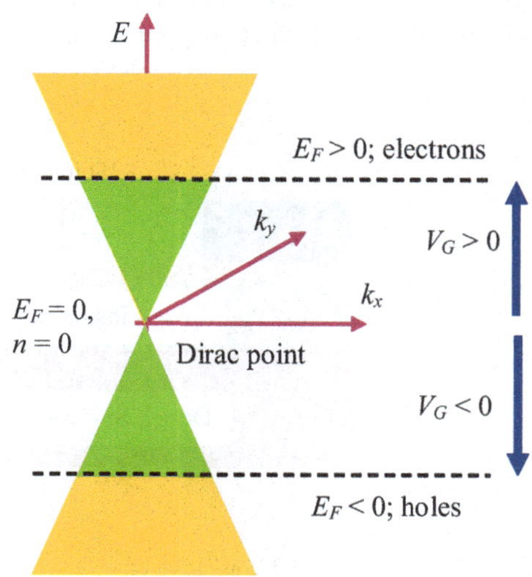

$$E(\mathbf{k}) = \pm\hbar|\mathbf{k}|v_{\mathrm{F}}, \tag{1.7}$$

where $v_F \cong c/300$ denotes the Fermi velocity, and $\mathbf{k} = (k_x, k_y)$ and $|\mathbf{k}| = [(k_x)^2 + (k_y)^2]^{1/2}$ are the wavenumber vector and its modulus, respectively.

In Fig. 1.24 we distinguish a neutrality point, termed Dirac point, where the density of carriers is zero. The Dirac point corresponds to the Fermi level of undoped graphene. The density of carriers can be increased with the help of a gate voltage. Depending on the orientation of this voltage, we have electrons or holes as carriers. This effect is similar to doping and has as consequence the shift of the Fermi energy level from the Dirac point. But, unlike in bulk semiconductors where the concentration of electrons and holes are determined by doping, in graphene the gate-induced doping is tunable and reversible. Moreover, in contrast to other semiconductors, the electrons and holes in graphene have identical physical properties and the transport is ambipolar due to the linearity of the dispersion relation.

Thus, the carrier type is selected by an electric field and the carrier density is given by (Novoselov et al. 2004):

$$n(V_{\mathrm{G}}) = \varepsilon_0 \varepsilon_r V_{\mathrm{G}}/te \tag{1.8}$$

where ε_0 is the dielectric permittivity of vacuum, ε_r the relative permittivity of the dielectric layer of the back gate, on which graphene is transferred (SiO_2, HfO_2), V_G is the back-gate voltage, t the thickness of the back-gate dielectric, and e the electric charge. The typical configuration of a back-gate graphene device is displayed in Fig. 1.25 and constitutes a field-effect transistor (FET).

Although a back-gate FET is not a performance transistor since large back-gate voltages are needed, it is widely used to study the physical properties of 2D materials. In the case of graphene, the Dirac point is determined from the drain current (I_D) versus gate voltage (V_G) dependence. This dependence has a V-like shape, the Dirac point being associated to the minimum current voltage, corresponding to 0 V in undoped devices. If the Dirac point is located at negative gate voltages, it means that

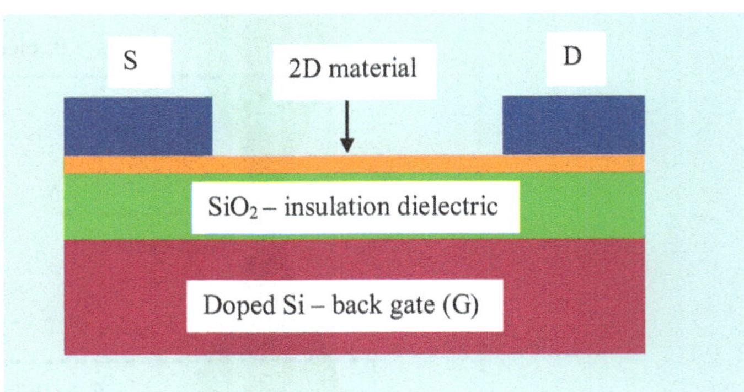

Fig. 1.25 Back-gate FET configuration

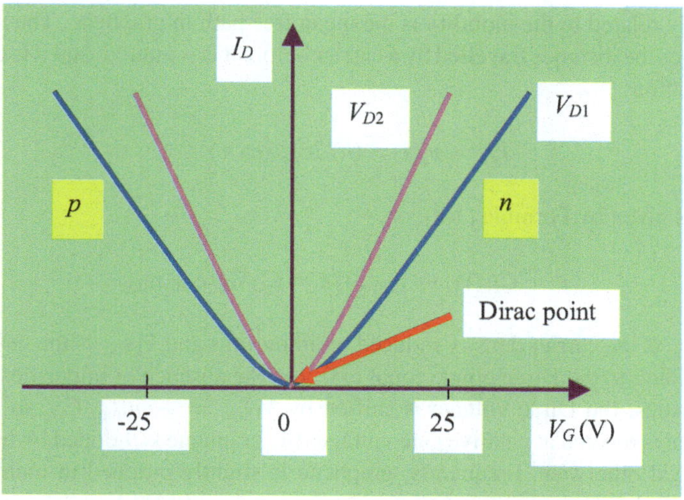

Fig. 1.26 Typical I_D–V_G dependence of a graphene monolayer in back-gate FET at two different drain voltages

the graphene is *p*-doped, while if it corresponds to positive voltages, it is *n*-doped (see Fig. 1.26). Only graphene shows such a distinctive I_D–V_G dependence, whereas for other semiconductor TMDs, with non-zero bandgaps, the dependence drain current versus gate voltage is similar to that of CMOS transistors.

The mobility is calculated as:

$$\mu = g_m L / W C_G |V_D| = g_m L t / \varepsilon W |V_D| \tag{1.9}$$

where $g_m = \partial I_D / \partial V_G$ is the transconductance, L and W are the length and the width of the FET channel, respectively, C_G is the gate capacitance per unit area, and V_D is the drain voltage. The values of carriers' mobility in graphene are displayed in Table 1.1. Table 1.1 indicates that graphene has a very large mobility at room temperature, but the scattering at the interface graphene/substrate plays a major role. If the graphene is not suspended, encapsulated with h-BN, or on a very low-roughness substrate, like ferroelectrics, e.g. if graphene is placed over a substrate such as SiO_2, the mobility attains only few thousand cm^2/V s.

Table 1.1 Mobilities in graphene

Mobility	Room temperature values (cm^2/V s)	References
Intrinsic	44,000	Shishir and Ferry (2009)
Suspended graphene or graphene over h-BN	100,000	Mayorov et al. (2011)
Graphene on ferroelectric	70,000	Hong et al. (2009)

Directly related to the mobility is the mean free path in graphene. The mean free path defines the distance travelled by a carrier before it is scattered, and it is expressed by the relation:

$$l_{fp} = v_F \tau = (h/2e)\mu(n/\pi)^{1/2} \tag{1.10}$$

where n is given by Formula (1.8), or

$$n = C_G(V_G - V_{Dirac})/e = C_G V_G/e + n_{imp} \tag{1.11}$$

where n_{imp} is the carrier density induced by impurities and V_{Dirac} is the gate voltage corresponding to the Dirac point (equal to 0 V in the absence of impurities). If there are impurities, the Dirac voltage is shifted by $V_{Dirac} = -en_{imp}/C_G$, so when the Dirac point is located at positive gate voltages the graphene is *p*-doped by impurities, and *n*-doped otherwise. Frequently, graphene is slightly *p*-doped in many devices due to the technological processing with resists such as PMMA.

The mean-free-path in graphene has very large values at room temperature, e.g. 400 nm if the graphene is deposited on SiO_2, beyond 1 μm when graphene is encapsulated in h-BN (Mayorov et al. 2011) and beyond 10 μm in graphene nanoribbons grown on SiC (Baringhaus et al. 2014). Table 1.2 presents various mean-free paths values in semiconductors and graphene for comparison. In devices with dimensions smaller than the mean-free-path the charge carrier transport is collisionless, this transport regime being denoted as ballistic.

These extremely large mean-free-path values are associated to wave-like ballistic carrier transport in graphene monolayers of very high quality, transport regime in which the linear graphene dispersion relation is equivalent to a vanishing effective mass of carriers:

$$m_{\text{eff},e} = m_{\text{eff},h} = 0 \tag{1.12}$$

Relation (1.12) follows from the dispersion relation in graphene and the definition of the effective mass $m_{\text{eff}} = (1/h^2)(\partial^2 E/\partial k^2)$. The ballistic transport regime confers also the highest possible cutoff frequencies of electronic devices. Otherwise, in the

Table 1.2 Room temperature mean free path in graphene and other semiconductors

Material	Mean free path
Si	3 nm
InAs nanowire	150 nm
InSb/AlInSb quantum well	580 nm
Graphene	500 nm in good quality graphene monolayer
	>1 μm for graphene encapsulated in BN monolayers
	>10 μm in graphene nanoribbons

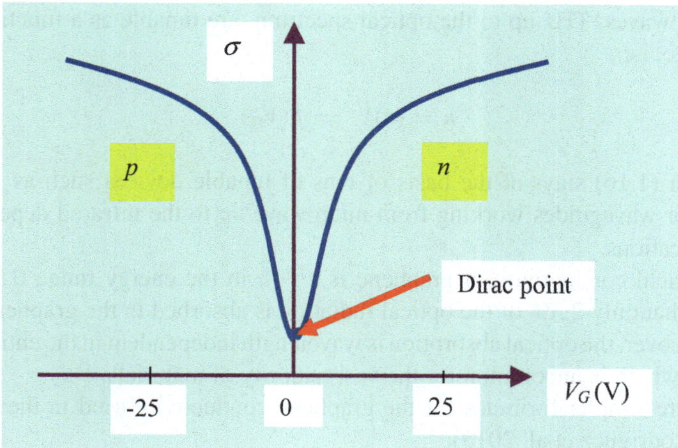

Fig. 1.27 Graphene conductance

presence of defects the transport equations are described by a drift-diffusion equation, like in other semiconductors. Therefore, the quality of graphene is so important.

The conductance of graphene, defined as $\sigma = I_D/V_D$, is displayed in Fig. 1.27. This parameter reveals many physical properties of graphene. The graphene conductance depends on frequency and is the sum of two conductances, due to intraband and interband transitions:

$$\sigma(\omega) = \sigma_1(\omega) + \sigma_2(\omega) \tag{1.13}$$

The graphene conductivity resulting from intraband transitions is given by:

$$\sigma_1(\omega) = ie^2 E_F/\pi\hbar^2(\omega + i/\tau) \tag{1.14}$$

while the conductivity contribution from interband transitions can be written as:

$$\sigma_2(\omega) = (ie^2\omega/\pi)\int_0^\infty [f(\mu_c - E_F) - f(-\mu_c - E_F)]/\left[(2\mu_c)^2 - (\hbar\omega + i\Gamma)\right]\mathrm{d}\varepsilon \tag{1.15}$$

where τ is the momentum relaxation time, μ_c is the chemical potential, Γ is a broadening parameter and $f(E)$ is the Fermi-Dirac distribution function. The range of frequencies where the conductivity is determined by intraband transitions is up to few THz, beyond which the conductivity due to interband transitions is dominating.

From formulas (1.14) and (1.15) it follows that both conductivities are determined by the position of the Fermi level. Since the permittivity or refractive index depend on the conductivity, it results that graphene devices working at high frequencies, from microwaves, THz up to the optical spectrum, are tunable as a function of the gate voltage, i.e.,

$$n = (\varepsilon)^{1/2} = f(V_{\mathrm{G}}) \tag{1.16}$$

Relation (1.16) stays at the basis of tens of tunable devices such as antennas, detectors or waveguides working from microwave up to the infrared depending on their applications.

The optical conductivity of graphene is $e^2/4\hbar$ in the energy range 0.5–1.2 eV, telling us that only 2.3% of the optical radiation is absorbed in the graphene monolayer. Moreover, the optical absorption is wavelength independent in the entire optical spectrum, which is uncommon for the vast majority of materials.

There are simpler formulas for the graphene conductivity used in the literature (Sensale-Rodriguez et al. 2013):

$$\sigma_1 = -i\frac{e^2 k_B T}{\pi\hbar^2(\omega - i2\Gamma)}\left(\frac{\mu_c}{k_B T} + 2\ln\left[\exp(-\mu_c/k_B T) + 1\right]\right) \tag{1.17a}$$

and

$$\sigma_2 = -i\left(\frac{e^2}{4\pi\hbar}\right)\ln\left(\frac{2|\mu_c| - \hbar(\omega - i2\Gamma)}{2|\mu_c| + \hbar(\omega - 2i\Gamma}\right). \tag{1.17b}$$

In addition, graphene shows unusual physical properties when excited by a magnetic field. Hall bars are used to study these effects, as illustrated in Fig. 1.28. When the magnetic field is applied perpendicular to the graphene plane, the energy becomes quantized on discrete Landau levels, given by:

$$E_n - V = \pm v_F\sqrt{2eB\hbar|n|} \tag{1.18}$$

with n an integer. Thus, the quantized values of the Hall conductivity are

$$\sigma_{xy} = (n + 1/2)(4e^2/h). \tag{1.19}$$

The quantized conductance value in (1.19) is dependent also on the gate voltage, which changes n. The Hall method is not only used to study graphene physical properties in magnetic field, but also its mobility via measurements of the resistivity or conductivity of graphene. More precisely,

$$\mu = (en\rho_{xy})^{-1} = \sigma_{xy}/en \tag{1.20}$$

Generally, the values of the carriers' mobility determined via Hall measurements are slightly different compared to those determined using FET DC measurements described above. However, both mobility values can be found in the literature.

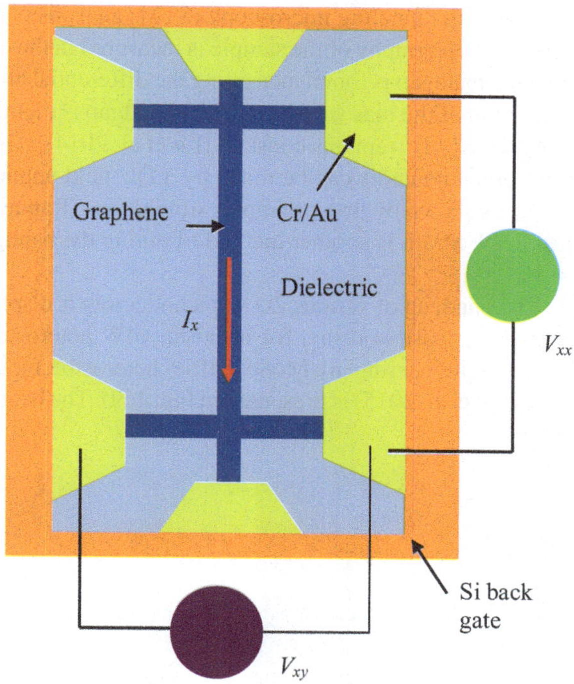

Fig. 1.28 Hall bar in graphene

Other 2D materials have different properties compared to graphene. For instance, TMDs are semiconductors, and the ballistic transport regime is not present at room temperature. In addition, their carrier transport is described by the Schrödinger equation or, more exactly, by the drift-diffusion equation. However, there are distinct physical properties of 2D TMDs compared to bulk semiconductors. Some of these properties of TMDs are summarized in Table 1.3.

The most interesting 2D materials are semiconductors such are TMDs. The bandgap is a basic parameter for any semiconductor. In atomically thin semiconductor materials, the bandgap is depending on the number of layers, a property not encountered in bulk semiconductors. For example, in MoS_2 the bandgap of a monolayer is 1.8 eV, of a bilayer is 2.10 eV, and that of a trilayer is 2.4 eV (Huang et al. 2015).

Table 1.3 Some physical properties of TMDs MX_2

M	*X*	Properties
Mo, W	S, Se, Te	X = S, Se—TMDs are semiconductors with large effective mass; e.g. WSe_2 $m_e = 0.33m_0/m_h = -0.46m_0$
Pd, Pt	S, Se, Te	X = S, Te—TMDs are semiconductors with a bandgap of 0.4 eV; tellurides are metallic

Scanning tunneling microscopy (STM) analysis is used to measure the bandgap. First, the topography of the sample is measured in the constant current mode regime, then the bandgap is found measuring the differential conductance spectra, dI/dV, as a function of the bias voltage, and the bandgap is identified from the applied voltage at which dI/dV remains constant (Lu et al. 2014). Also, the bandgap of atomically thin semiconductors can be tuned by a DC field applied vertically to the material's plane, as shown by first-principles simulations (Ramasubramaniam et al. 2011); see Fig. 1.29. Strain is another method of tuning the bandgap of 2D semiconductors up to 5%.

The bandgap of various 2D semiconductors is displayed in Table 1.4. Band engineering is possible using, for instance, vdW heterostructures. An example of such bandgap engineering in a $MoSe_2/HfSe_2$ heterostructure grown on AlN(0001)/Si(111) (Aretouli et al. 2015) is presented in Fig. 1.30. The heterostructure consists of several

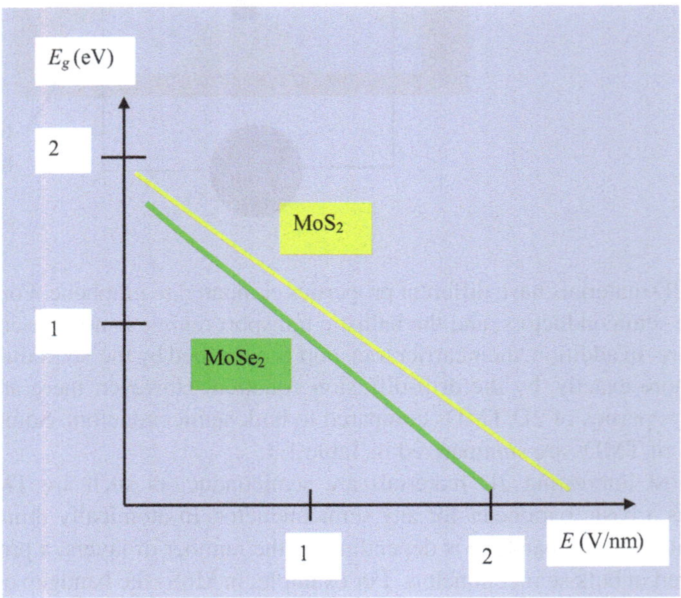

Fig. 1.29 Tunning the bandgap of 2D semiconductors by a vertical DC field

Table 1.4 2D materials' bandgap (Akinwande et al. 2014; Kanazawa et al. 2016)

2D material	Bandgap (eV)	Bandgap type	Material type
Graphene	0	Direct	Semimetal
MoS_2 monolayer	1.8	Direct	Semiconductor
HfS_2 monolayer	1.2	Direct	Semiconductor
Phosphorene	0.5–2	Direct	Semiconductor
h-BN	5.9	Direct	Insulator

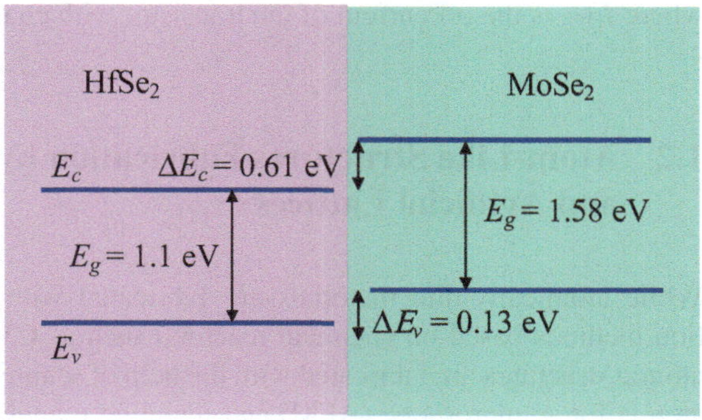

Fig. 1.30 Type II $MoSe_2/HfSe_2$ heterostructure

$MoSe_2$ layers grown on top of $HfSe_2$, which in turn is grown on AlN. A more detailed analysis of bandgap in 2D semiconductors, and their properties are found in (Dragoman and Dragoman 2017).

The mobility and effective masses are essential parameters for circuits based on 2D semiconductors, and most of these parameters are unknown from experimental results. Therefore, the only way to estimate them is through atomistic simulations (Zhang et al. 2014). However, although the mobility is originating from scattering mechanisms involving phonons, impurities, electrons, etc., in simulations only acoustic phonon scattering was considered, e.g. only the first term from the formula below

$$1/\mu = 1/\mu_{\text{photons}} + 1/\mu_{\text{impurity}} + 1/\mu_{\text{electron}} + \cdots \tag{1.21}$$

In Table 1.5, we have displayed the mobility and the effective mass of few 2D semiconductors. From Table 1.5 it follows that the mobility of 2D semiconductors such as MoS_2 and $MoSe_2$ is lower than that of Si, whereas the mobility of 2D HfS_2 and $HfSe_2$ is much higher than this value, but these later values are not confirmed experimentally yet. However, almost all 2D semiconductors have high and very anisotropic effective masses. This fact is used as an argument to show that transistors can be downscaled up to 1 nm (Ilatikhameneh et al. 2016), i.e., up to the dimensions of a few atoms, since the leakage current is proportional to the effective mass:

$$\log(I_{\text{OFF}}) = -L_{\text{ch}}\sqrt{m_{\text{eff}}} \tag{1.22}$$

Table 1.5 2D materials' mobility and effective masses (Zhang et al. 2014)

2D material	Mobility (cm^2/V s)	$m_{\text{eff}}\ \Gamma - K \times m_e$	$m_{\text{eff}}\ K - M \times m_e$
MoS_2	340	0.45	0.45
$MoSe_2$	240	0.52	0.52
HfS_2	1833	3.30	0.24
$HfSe_2$	3579	3.10	0.18

where I_{OFF} is the off current of the transistor with channel length L_{ch}.

1.2 Atom-Like Structures Fabrication by Bonds, Dopants and Artificial Lattices

While atomically-thin materials are fabricated with the help of various deposition methods based on chemical reactors, such as CVD, ALD, MBE, the ultimate atomic structures are fabricated with the help of scanning probe equipments, such as atomic force microscope (AFM) and scanning tunneling microscope (STM) at low temperatures and under special pressure conditions.

Since the principles of AFM and STM are explained extensively in various places for different applications, as for example in Dragoman and Dragoman (2009b), we will present them below only briefly to support the understanding of fabrication at the atomic scale.

The AFM is able to detect and manipulate atoms and molecules. It is based on the very weak forces, even of attonewton level ($1\ \text{aN} = 10^{-18}$ N), exerted between the cantilever and the atoms or molecules on a sample's surface (Bhushan 2004). These forces applied between the surface and the cantilever, terminated with a tip which has a very small mass, induce the deflection of the cantilever, determined by optical interferometry. The cantilever could be functionalized or even biased to act selectively/in a desired way on atoms or molecules.

Depending on the distance between the tip of the cantilever and the sample, the AFM is able to sense various types of forces, which in turn dictate the interaction between the tip and the sample. At very small tip-sample surface distances z (for $z \leq 0.1$ nm), we are dealing with Born repulsive force, given by

$$U_{\text{Born}} = C_1/z^{12} \tag{1.24a}$$

where C_1 is a constant.

At larger tip-surface distances, below 10 nm, the van der Waals force is dominating, whereas at still larger distances other forces could act, such as electrostatic and/or magnetic forces. The vdW force is expressed as

$$U_{\text{vdW}} \approx -C_2/z^6, \tag{1.24b}$$

In the above formula, the constant C_2 is the London coefficient.

In general, the total potential between the tip and the surface is the sum of the Born and van der Waals potentials:

$$U = C_2/z^{12} - C_1/z^6 \tag{1.25}$$

the above expression being known as the Lenard-Jones potential.

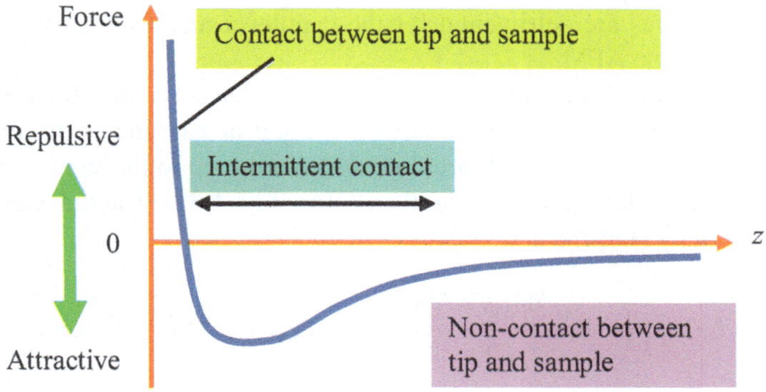

Fig. 1.31 Operation modes of an AFM

The force detected with the AFM is the sum of all repulsive and attractive forces which are acting during the tip-sample interaction. The dependence of the force $F = -\partial U/\partial z$ on z determines the three types of AFM modes of operation, represented in Fig. 1.31: contact mode, intermittent contact, and non-contact mode.

In what follows, of interest is the *dynamical* or the *non-contact* mode. In this mode the cantilever tip is not in contact with the surface under test, being placed always above the surface and scanning it. The cantilever is a mechanical oscillator, oscillating at a determined frequency f_0 with the help of a piezo-actuator. The reduction of the noise and other adverse effects is achieved by modulating the cantilever motion via amplitude modulation (AM) or frequency of modulation (FM).

The principle of optical detection of the cantilever motion is represented in Fig. 1.32. The position of the mechanically deflected cantilever is sensed by an array of four detectors, each of them detecting the current I_i where $i = 1,\ldots, 4$. These currents are proportional to the forces exerted on the cantilever:

$$\Delta_{\text{normal force}} \propto [(I_1 + I_2) - (I_3 + I_4)] \tag{1.26a}$$

$$\Delta_{\text{lateral force}} \propto [(I_2 + I_4) - (I_1 + I_3)] \tag{1.26b}$$

Fig. 1.32 AFM optical signal detection

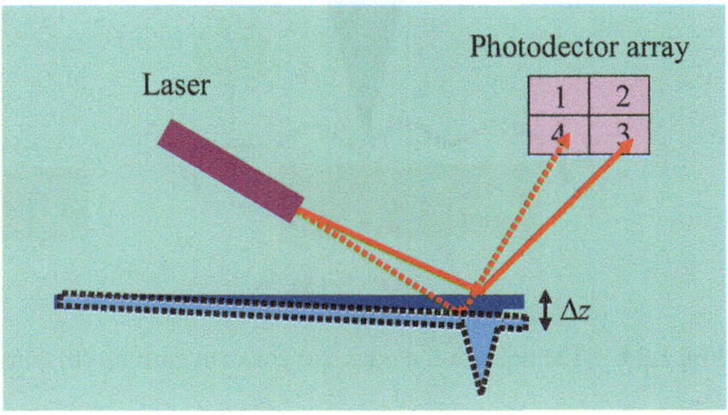

The lateral forces are detrimental for the cantilever and are compensated by a feedback loop inside AFM.

The scanning tunneling microscope (STM) is based on the tunneling effect between a sharp metallic tip and a surface located at nanometer distances from the tip. The electrons of energy E are flowing from the tip to the sample, tunneling a vacuum barrier with height ϕ greater than E (see Fig. 1.33). The tunneling current density is given by Hansma and Tersoff (1987)

$$i = V\rho_{\mathrm{surf}}(E_{\mathrm{F}})\exp\left(-2\frac{\sqrt{2m(\phi - E)}}{\hbar}z\right) \propto V\rho_{\mathrm{surf}}(E_{\mathrm{F}})\exp\left(-1.025\sqrt{\phi}z\right) \quad (1.27)$$

where we have denoted by $\rho_{\mathrm{surf}}(E_{\mathrm{F}})$ the density of states (DOS) of the surface, z is the distance between the tip and the surface, i.e., the vacuum barrier width, and m is the electron mass.

The STM can operate in two modes: (i) The constant current mode, in which the tunneling current is monitored when the tip is located near the surface and is biased. The current is kept constant by a feedback loop which changes z in this scope and memorizes it. (ii) The constant height mode, in which the tip scans the surface at a constant height and the current is directly monitored, measured and controlled allowing the feedback loop to preserve a constant height. The two operating modes are displayed in Fig. 1.34. There are various STM equipments and systems that could

Fig. 1.33 The STM physical principle

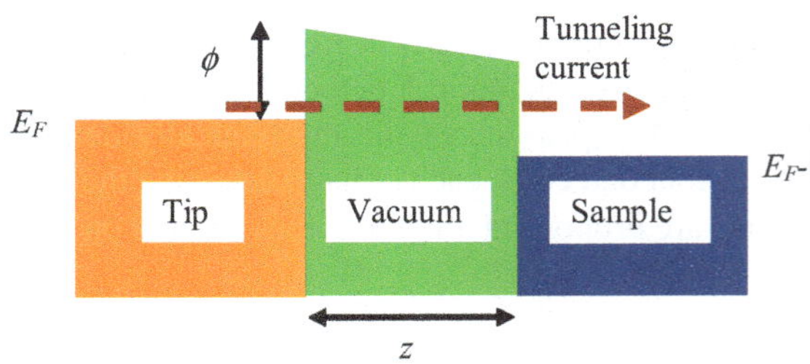

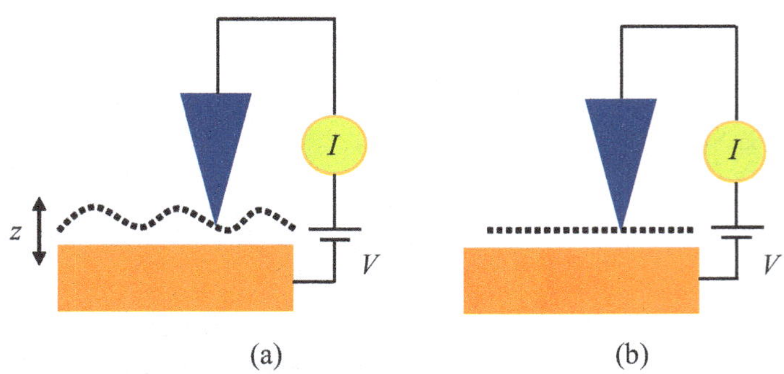

Fig. 1.34 STM operation modes: (a) constant current, (b) constant height

operate at low temperatures (350 mK), high magnetic fields (11 T) and ultrahigh vacuum (UHV).

Both AFM and STM are used intensively in nanofabrication processes (Cui 2008). For example, the STM is used for resist exposure in an analogous manner to e-beam lithography, since electrons are emitted from the tip in the vicinity of the resist and the STM is working in the constant current mode. Features well below 50 nm can be obtained in this way, which are rather difficult to fabricate with e-beam lithography.

Another example is local oxidation, which can be performed with the AFM or the STM. If only few volts are applied on the tip placed at few nanometers from the surface, an electric field of 10^9 V/cm is produced, and at this high field the molecules are dissociated into ions. For instance, Si can be passivated in HF such that a single atomic layer of H remains on Si, termed also Si hydrogen (H) terminated. The local oxidation of this material occurs when water molecules are present around the tip. The hydrogen can thus be locally removed with the STM and the Si surface is oxidated, becoming SiO_2. Deposition and etching are also possible with scanning probe equipments.

Following the above techniques, one of the most promising methods of nanofabrication is hydrogen lithography, where a single H atom is removed from a H passivated Si surface, producing a single dangling-bond (DB), when a STM is placed above the surface and a voltage is applied between the tip and the sample (see Fig. 1.35) (Pavlicĕk et al. 2017).

The DBs behave like quantum dots, i.e., as artificial atoms having a discrete energy level spectrum. The DBs can be seen also as adsorption sites for single dopants such as phosphine, which, after activation, is transformed into a phosphorous dopant.

The most important issue regarding DBs with atomic precision placement is the functionalization of the tip. In this respect, two kinds of tips are necessary: a chemical inert tip, which is a H tip, and a reactive tip, which is a Si tip (Pavlicĕk et al. 2017).

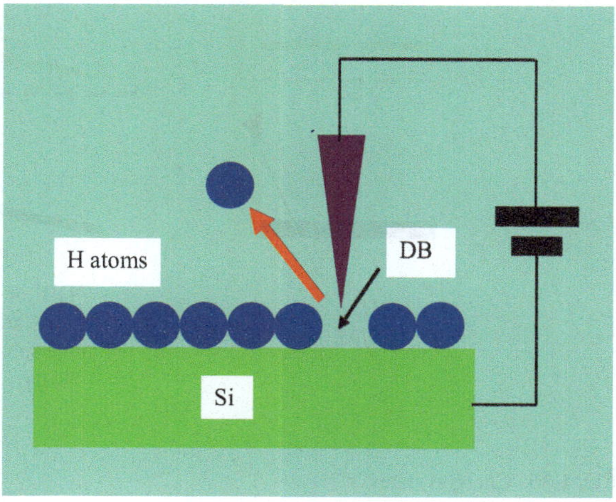

Fig. 1.35 Principle of hydrogen lithography

The Si tip is used to write a DB (hydrogen desorption), while the H tip is used for error correction by hydrogen passivation.

The DBs were obtained working with STM at 4 K, STM and AFM images being recorded to validate the quality of the process. The DB appears in the STM image as a depression, i.e., as a dark disk having 2 nm lateral extension and a depth of 150 pm. The DB is in fact a defect having a bound state below the conduction band, and this state can be occupied by 0, 1, or 2 electrons, making it equivalent to a positive, neutral or negative charge state (Schofield et al. 2013).

Thus, we can "write" artificial molecules and even one- and two-dimensional solids using a Si(001) surface passivated with hydrogen. With atomic precision we can remove hydrogen atoms from the surface with STM at low temperatures. The STM can work with negative and positive DC voltages, inducing, respectively, filled and empty states, each having a distinct STM image. In the filled state, a negatively charged DB is detected, while in the empty state a positively or a negatively charged DB is detected by STM when an equilibrium state is reached, or not, respectively, with the substrate (see Fig. 1.36).

Thus, a DB with no electrons is positively charged and denoted DB^+, an electron can transform this DB in a neutral state denoted DB^0, and the addition of another electron is transforming the DB in a negatively charged state denoted DB^-. STM observes the transitions (+/0) and (0/−), which manifest themselves as steps in the current (I)–voltage (V) dependence. However, there is a clear feature in the $I-V$ dependence when the current is decreasing while the voltage is increasing. This phenomenon is termed as negative differential resistance (NDR) (see Fig. 1.37) and is widely spread in electronics, being the key concept of any oscillator (Dragoman and Dragoman 2009a).

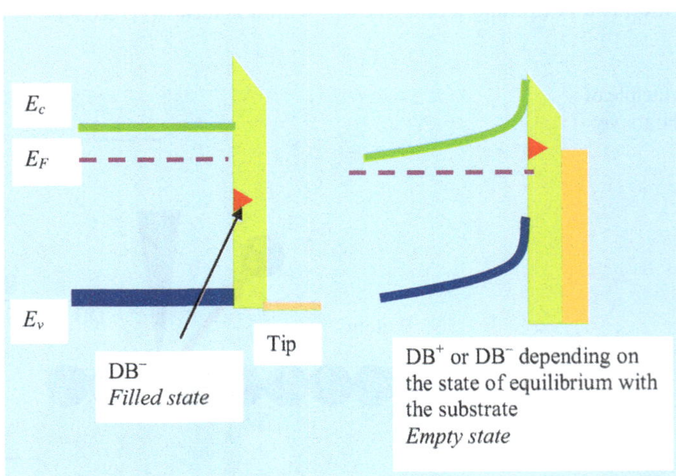

Fig. 1.36 Charged states of DB

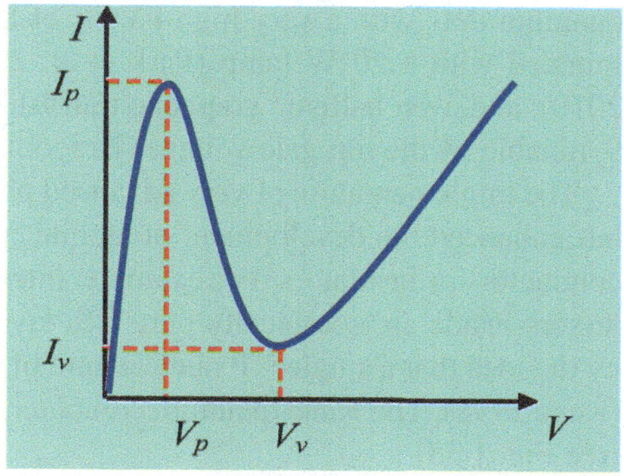

Fig. 1.37 The NDR

In the case of DB, it was observed by Rashidi et al. (2016) with the help of a STM at 4.5 K. The NDR is launched when the tip energy is resonant with the (+/0) level, which becomes empty, and electrons are flowing from the conduction band through the (+/0) level, so that (+/0) is acting as an energy filter for electrons and allows the flow of only those electrons with the same energy as the (+/0) state. This is a similar process to a resonant tunneling diode, where electrons propagate between the two barriers only at energies equal to the quantum well resonance level.

From the NDR curve we can determine several parameters. The first and the second positive differential resistances are given by

$$R_{p1} = V_p / I_p, \tag{1.28a}$$

$$R_{p2} = (V_v - V_p)/(I_p - I_v), \tag{1.28b}$$

the NDR is expressed by

$$R_n = (V_p - V_v)/(I_p - I_v), \tag{1.29}$$

whereas the peak-to-valley voltage ratio and the peak-to-valley current ratio are defined as

$$\text{PVVR} = V_p / V_v \tag{1.30}$$

$$\text{PVCR} = I_p / I_v. \tag{1.31}$$

In the case of NDR in DBs, the NDR has the following parameters collected from Fig. 1(a) in Rashidi et al. (2016) $I_p = 1200\,\text{pA}$, $I_v = 200\,\text{pA}$, so the PVCR = 6, $V_p = -1.1\,\text{V}$, $V_v = -1.4\,\text{V}$, and PVVR = 0.78.

We note that the NDR is observed in many nanostructures, including molecular devices (Deng et al. 2009), ZnO nanobelts (Ya et al. 2010), room temperature quantum dots with a very high PVCR of 80 in the dark state and 2240 when illuminated with a 20 W lamp (Park et al. 2006), GaN nanowires (Dragoman et al. 2010), and even ballistic graphene transistors at room temperature with a PVCR of 9, tunable by the top-gate voltage (Dragoman et al. 2014).

The implementation of very advanced physical ideas such as quantum computing has enhanced the development of atomic control placement of defects inside semiconductors or insulators. For example, intensive efforts to place single phosphorous donors inside an isotopically pure ^{28}Si layer (Itoh and Watanabe 2014) are fuelled by the idea that a single ^{31}P nuclear spin of each P donor can be used as a spin qubit (Kane 1998). The Kane quantum computer has a back gate and several control gates (see Fig. 1.38)

The ^{31}P atoms have nuclear spin ½ while the rest of ^{28}Si atoms have nuclear spin 0. The magnetic field B is used to manipulate the spins (the two qubits in the two defects in Fig. 1.38). The gates G_S are used to address each spin qubit by changing their Larmour frequency, while the G_C gate is used to couple them via electron-mediated interaction between nuclear spins. If we want more qubits, we need to introduce more defects in Si.

The dopants have as DBs three states: an ionized state, denoted D^+, which is an empty state that does not reside in the potential energy well felt by the dopant in the host material, a neutral state D^0 where one electron is bound to the dopant, and a

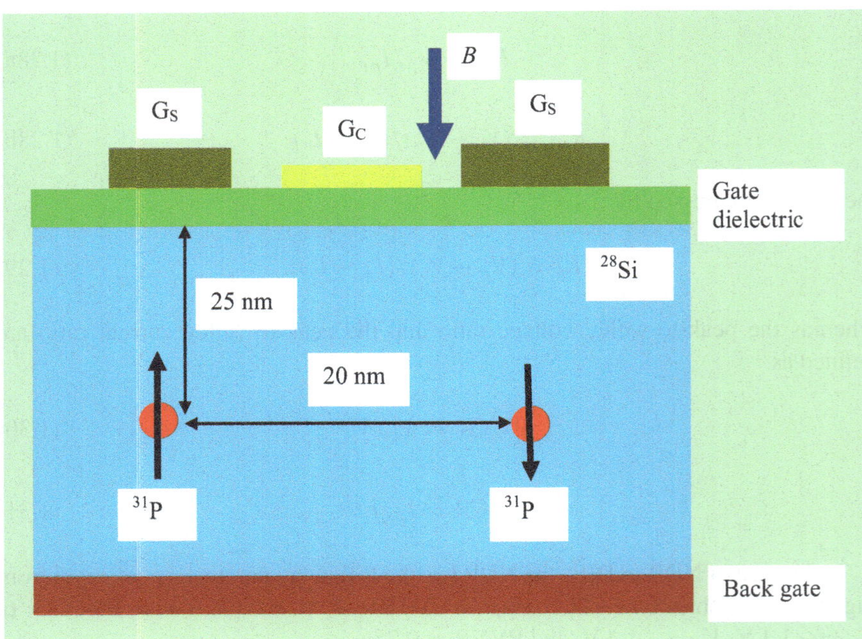

Fig. 1.38 Phosphorus defects for quantum computing

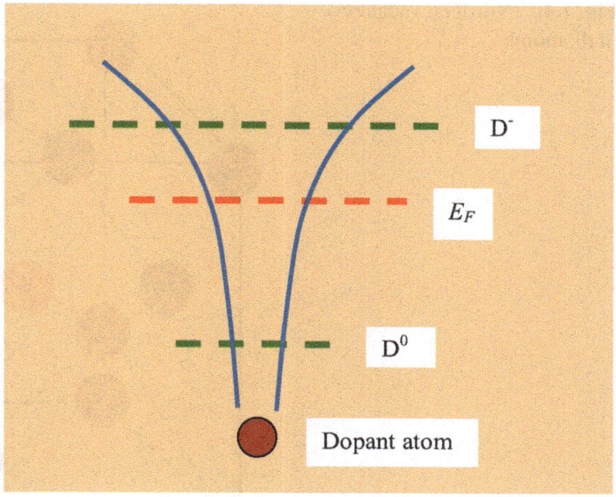

Fig. 1.39 The potential confinement of a single dopant

negative charge state D^-, where two electrons are bound to the dopant (Zwanenburg et al. 2013). The confinement potential of a dopant is represented in Fig. 1.39.

The applications of single dopants will be developed in the chapter dedicated to atomic electronics. Here we note that it is rather difficult to get a high yield using STM to introduce dopants, therefore industrial methods are used to perform this thorny task by ion implantation of P, As, Sb, or Bi. Post-implantation techniques are then used to locate the best places for control and readout via tuned gate voltages. In particular, the 14 keV ^{31}P donors are successfully implanted in Si via deterministic ion implantation (van Donkelaar et al. 2015).

Similar concepts for quantum computation use nitrogen vacancies (NV) in diamond, which are also defects inside the perfect crystalline structure of diamond. The NV in diamond is an atom-like structure acting as dopant. A NV center is a point defect formed from a nitrogen atom replacing a carbon atom and having a lattice vacancy in the vicinity (see Fig. 1.40). The NV has a negative charge configuration where it has a spin $S = 1$ on the vacancy and its resonance can be reached with light excitation at 637 nm. The spin of NVs can be prepared and detected using the optical transitions of NV, and based on this the first quantum gates and algorithms working at room temperature were experimentally proven (Doherty et al. 2013). The NV centers are native or are created by ion implantation.

Atom-like structures can be obtained at much larger scale. For example, graphene, which consists of a hexagonal atom lattice (see Fig. 1.41) is mimicked by an artificial graphene (AG) with the same lattice, where different physical phenomena associated to graphene can be studied at much larger scales, for example in hexagonal lattices working at microwaves or in photonics. The graphene monolayer consists of a sheet of carbon atoms hybridized in the sp^2 state. Each C atom form a covalent bond with three other atoms, and thus a honeycomb lattice is formed from two interpenetrating triangular sublattices. The colours in Fig. 1.41, which emphasize the two triangular sublattices, are only for visualisation purposes; all atoms are identical carbon atoms.

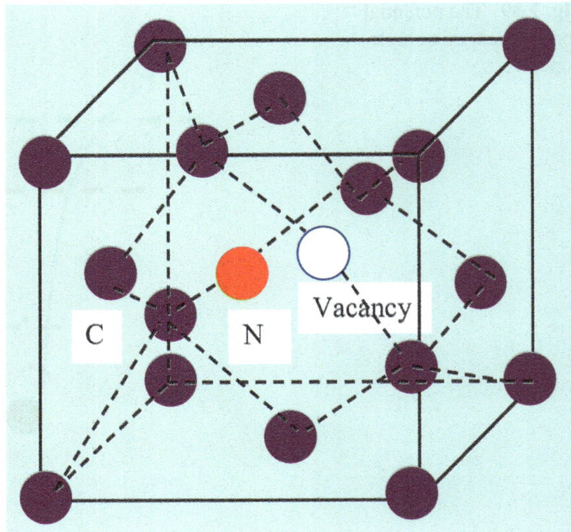

Fig. 1.40 Nitrogen vacancy in diamond

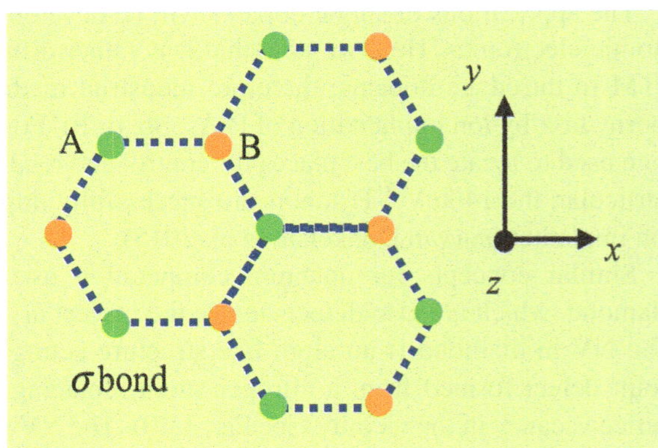

Fig. 1.41 The graphene lattice

Although not all physical phenomena occurring in graphene can be studied by an AG, both structures, atomic or at larger scale, have the same dispersion relation due to their shape—the hexagonal lattice. Artificial defects, strong-orbital coupling, or topological phases are studied in such artificial honeycomb lattices which can be implemented by nanopatterning of 2D gas heterostructures, molecular assembly by STM, cold atoms, or microwave and photonics structures (Polini et al. 2013).

For instance, a microwave analogue to graphene is made from metallic cylindrical rods with diameter of 2 mm and length of 15 mm arranged in a honeycomb lattice with a period of 5 mm. The entire area is 2.5×3 cm^2, measurements being made in the range 1–25 GHz. The dispersion curve consists of two crossing lines, as in the case of graphene monolayers (Dautova et al. 2017). In another example, using a strained honeycomb photonic lattice, pseudomagnetic fields and photonic Landau levels in dielectric structures were revealed (Rectsman et al. 2012), while Floquet topological insulators were evidenced in Rectsman et al. (2013).

References

Akinwande D, Petrone N, Hone J (2014) Two-dimensional flexible nanoelectronics. Nat Commun 5:5678

Allain A, Kang J, Banerjee K, Kiss A (2015) Electrical contacts to two-dimensional semiconductors. Nat Mater 14:1195–1205

Aretouli KE, Tsipas P, Tsoutsou D, Marquez-Velasco J, Xenogiannopoulou Giamini SA, Vassalou E, Kelaidis N, Dimoulas A (2015) Two-dimensional semiconductor HfSe and MoSe2/HfSe2 van der Waals heterostructures by molecular beam epitaxy. Appl Phys Lett 106:143105

Baringhaus J, Ruan M, Edler F, Tejeda A, Sicot M, Taleb-Ibrahimi A, Li A-P, Jiang Z, Conrad EH, Berger C, Tegenkamp C, de Heer WA (2014) Exceptional ballistic transport in epitaxial graphene nanoribbons. Nature 506:349–354

Bhushan B (ed) (2004) Springer handbook of nanotechnology. Springer

Cheng Z, Price K, Franklin AD (2018) Contacting and gating 2-D nanomaterials. IEEE Trans Electron Dev 65:4073–4083

Cho Y, Berkelbach TC (2018) Environmentally sensitive theory of electronic and optical transitions in atomically thin semiconductors. Phys Rev B 97:041409(R)

Cui Z (2008) Nanofabrication principles, capabilities and limits. Springer, New York

D'Angelo M, Matsuda I (2019) Basics and families of monoatomic layers: single-layer 2D materials. In: Matsuda I (ed) Monoatomic two-dimensional layers. Modern experimental approaches for structure, properties and industrial use. Elsevier, pp 3–22

Dautova YN, Shytov AV, Hooper IR, Sambles JR, Hibbins AP (2017) Gapless states in microwave artificial graphene. Appl Phys Lett 110:261605

de Heer WA, Berger C, Ruan M, Sprinkle M, Li X, Hu Y, Zhang B, Hankinson CJE (2011) Large area and structured epitaxial graphene produced by confinement controlled sublimation of silicon carbide. PNAS 108:16900–16905

Deng XQ, Zhou JC, Zhang ZH, Zhang H, Qiu M (2009) Electrode conformation-induced negative differential resistance and rectifying performances in a molecular device. Appl Phys Lett 95:163109

Doherty MW, Mansonb NB, Delaney P, Jelezko F, Wrachtrupe J, Hollenberg LCL (2013) The nitrogen-vacancy colour centre in diamond. Phys Rep 528:1–45

Dragoman M, Dragoman D (2009a) Graphene-based quantum electronics. Progr Quantum Electron 33:165–214

Dragoman M, Dragoman D (2009b) Nanoelectronics: principles and devices (2nd edn). Artech House

Dragoman M, Konstantinidis G, Cismaru A, Vasilache D, Dinescu A, Dragoman D, Neculoiu D, Buiculescu R, Deligiorgis G, Vajpeyi AP, Georgakilas A (2010) Negative differential resistance in GaN nanowire network. Appl Phys Lett 96:053119

Dragoman M, Dinescu A, Dragoman D (2014) Negative differential resistance in graphene based ballistic field-effect transistor with oblique top gate. Nanotechnology 25:415201

Dragoman M, Dragoman D (2017) 2D nanoelectronics. Physics and devices of atomically thin materials. Springer, Cham

Duong DL, Yun SJ, Lee YH (2017) van der Waals layered materials: opportunities and challenges. ACS Nano 11:11803–11830

Frank O, Kavan L, Kalbac M (2014) Carbon isotope labelling in graphene research. Nanoscale 6:6363

Fujita J-I, Hiyama T, Hirukawa A, Kondo T, Nakamura J, Ito S-I, Araki R, Ito Y, Takeguchi M, Pai WW (2017) Near room temperature chemical vapor deposition of graphene with diluted methane and molten gallium catalyst. Sci Rep 7:12371

Geim AK, Grigorieva IV (2013) Van der Waals heterostructures. Nature 499:419–425

Geng D, Yang HY (2018) Recent advances in growth of novel 2D materials: beyond graphene and transition metal dichalcogenides. Adv Mater 30:1800865

Gong Y, Lin J, Wang X, Shi G, Lei S, Lin Z, Zou X, Ye G, Vajtai R, Yakobson BI, Terrones H, Terrones M, Tay BK, Lou J, Pantelides ST, Liu Z, Zhou W, Ajayan PM (2014) Vertical and in-plane heterostructures from WS_2/MoS_2 monolayers. Nat Mater 13:1135–1142

Ilatikhameneh H, Ameen T, Novakovic B, Tan Y, Klimeck G, Rahman R (2016) Saving Moore's law down to 1 nm channels with anisotropic effective mass. Sci Rep 6:301501

Jang A-R, Hong S, Hyun C, Yoon SI, Kim G, Jeong HY, Shin TJ, Park SO, Wong K, Kwak SK, Park N, Yu K, Choi E, Mishchenko A, Withers F, Novoselov KS, Lim H, Shin HS (2016) Wafer-scale and wrinkle-free epitaxial growth of single-orientated multilayer hexagonal boron nitride on sapphire. Nano Lett 16:3360–3366

Jin Z, Shin S, Kwon DH, Han SH, Min YS (2014) Novel chemical route for atomic layer deposition of MoS_2 thin film on SiO/Si substrate. Nanoscale 6:14453–14458

Jung Y, Ji E, Capasso A, Lee G-H (2019) Recent progresses in the growth of two-dimensional transition metal dichalcogenides. J Korean Ceramic Soc 56:24–36

Hansma PK, Tersoff J (1987) Scanning tunneling microscope. J Appl Phys 61:R1–R23

Hao W, Marichy C, Journe C (2019) Atomic layer deposition of stable 2D materials. 2D materials 6:012001

Hong X, Posadas A, Zou K, Ahn CH, Zhu J (2009) High-mobility few-layer graphene field effect transistors fabricated on epitaxial ferroelectric gate oxides. Phys Rev Lett 102:136808

Huang YL, Chen Y, Zhang W, Quek SY, Chen C-H, Li L-J, Hsu W-T, Chang W-H, Zheng YJ, Chen W, Wee ATS (2015) Bandgap tunability at single-layer molybdenum disulphide grain boundaries. Nat Commun 6:6298

Itoh KM, Watanabe H (2014) Isotope engineering of silicon and diamond for quantum computing and sensing applications. MRS Commun 4:143–157

Kanazawa T, Amemiya T, Ishikawa A, Upadhyaya V, Tsuruta K, Tanaka T, Miyamoto Y (2016) Few-layer HfS_2 transistors. Sci Rep 6:22277

Kane BE (1998) A silicon-based nuclear spin quantum computer. Nature 393:133–137

Kang K, Xie S, Huang L, Han Y, Huang PY, Mak KF, Kim C-J, Muller D, Park J (2015) High-mobility three-atom-thick semiconducting films with wafer-scale homogeneity. Nature 520:656–660

Kaur G, Kavitha K, Lahiri I (2019) Transfer-free graphene growth on dielectric substrates: a review of the growth mechanism. Crit Rev Solid State Mater Sci 44:157–209

Koma A, Sanouchi K, Miyajima T (1984) Fabrication and characterization of heterostructures with subnanometer thickness. Microelectron Eng 2:129–136

Kwak T, Lee J, So B, Choi U, Nam O (2019) Growth behavior of wafer-scale two-dimensional MoS_2 layer growth using metal-organic chemical vapor deposition. J Crystal Growth 510:50–55

Lee Y-H, Zhang X-Q, Zhang W, Chang M-T, Lin C-T, Chang K-D, Yu Y-C, Wang JT-W, Chang C-S, Li L-J, Lin T-W (2012) Synthesis of large-area MoS_2 atomic layers with chemical vapor deposition. Adv Mater 24:2320–2325

Liu T, Liu S, Kun-Hua Tu K-H, Schmidt H, Chu L, Xiang D, Martin J, Eda G, Ross CA, Garaj S (2019) Crested two-dimensional transistors. Nat Nanotechnol 14:223–226

Lu C-P, Li G, Mao J, Wang L-M, Andrei EY (2014) Bandgap, mid-gap states, and gating effects in MoS_2. Nano Lett 14:4628–4633

Mayorov AS, Gorbachev RV, Morozov SV, Britnell L, Jail R, Ponomarenko LA, Novoselov KS, Watanabe K, Taniguchi T, Geim AK (2011) Micrometer-scale ballistic transport in encapsulated graphene at room temperature. Nano Lett 11:2396–2399

Miró P, Audiffred M, Heine T (2014) An atlas of two-dimensional materials. Chem Soc Rev 43:6537–6554

Mishra N, Boeckl J, Motta N, Iacopi F (2016) Graphene growth on silicon carbide: A review. PSS 213A:2277–2289

Novoselov KS, Geim AK, Morozov SV, Jiang D, Zhang Y, Dubonos SV, Grigorieva IV, Firsov AA (2004) Electric field effect in atomically thin carbon films. Science 306:666–669

Park NM, Kim SH, Maeng S, Park S-J (2006) Negative differential resistance in silicon quantum dot metal-insulator semiconductor structure. Appl Phys Lett 89:153117

Pavliček N, Majzik Z, Meyer G, Gross L (2017) Tip-induced passivation of dangling bonds on hydrogenated Si(100)-2 $\times$ 1. Appl Phys Lett 111:053104

Polini M, Guinea F, Lewenstein M, Manoharan HC, Pellegrini V (2013) Artificial honeycomb lattices for electrons, atoms and photons. Nat Nanotechnol 8:625–633

Ramasubramaniam A, Naveh D, Towe E (2011) Tunable band gaps in bilayer transition-metal dichalcogenides. Phys Rev B 84:205325

Rashidi M, Taucer M, Ozfidan I, Lloyd E, Koleini M, Labidi H, Pitters J, Maciejko J, Wolkow RA (2016) Time-resolved imaging of negative differential resistance at atomic scale. Phys Rev Lett 117:276805

Rashidi M, Lloyd E, Huff TR, Achal R, Taucer M, Croshaw JJ, Wolkow RA (2017) Resolving and tuning carrier capture rates at a single silicon atom gap state. ACS Nano 11:11732–11738

Rashidi M, Taucer M, Ozfidan I, Lloyd E, Koleini M, Labidi H, Pitters JL, Maciejko J, Wolkow RA (2016) Time-Resolved Imaging of Negative Differential Resistance on the Atomic ScalePhys. Rev. Lett. : 117, 276805

Rechtsman MC, Zeuner JM, Tünnermann A, Nolte S, Segev M, Szameit A (2012) Strain-induced pseudomagnetic field and photonic Landau levels in dielectric structures. Nat Photonics 7:153–158

Rechtsman MC, Zeuner JM, Plotnik Y, Lumer Y, Podolsky D, Dreisow F, Nolte S, Segev M, Szameit A (2013) Photonic Floquet topological insulators. Nature 496:196–200

Schofield SR, Studer P, Hirjibehedin CF, Curson NJ, Aeppli G, Bowler DR (2013) Quantum engineering at silicon surface using dangling bonds. Nat Commun 4:1649

Schulman DS, Arnolds AJ, Das S (2018) Contact engineering for 2D materials and devices. Chem Soc Rev 47:3037–3058

Schwierz F, Pezoldt J, Grazner R (2015) Two-dimensional materials and their prospects in transistor electronics. Nanoscale 7:8261–8283

Sensale-Rodriguez B, Yan R, Liu L, Jena D, Xing HG (2013) Graphene for reconfigurable terahertz optoelectronics. Proc IEEE 101:1705–1716

Shishir RS, Ferry DK (2009) Intrinsic mobility in graphene. J Phys: Cond Matter 21:232204

Song X, Gao J, Nie Y, Gao T, Sun J, Ma D, Li Q, Chen Y, Jin C, Bachmatiuk A, Rümmeli MH, Ding F, Zhang Y, Liu Z (2015) Chemical vapor deposition growth of large-scale hexagonal boron nitride with controllable orientation. Nano Res 10:3164–3176

Sundaram S, Li X, Alam S, Halfaya Y, Patriarche G, Ougazzaden A (2019) Wafer-scale MOVPE growth and characterization of highly ordered h-BN on patterned sapphire substrates. Cristal Growth 509:40–43

Tai G, Hu T, Zhou Y, Wang X, Kong J, Zeng T, You Y, Wang Q (2015) Synthesis of atomically thin boron films on copper foils. Angew Chem Int Ed 54:15473–15477

Tongay S, Fan W, Kang J, Park J, Koldemir U, Suh J, Narang DS, Liu K, Ji J, Sinclair R, Wu J (2014) Tuning interlayer coupling in the large-area heterostructures with CVD-grown MoS_2 and WS_2 monolayers. Nano Lett 14:3185–3190

Tosun M, Fu D, Desai SB, Ko C, Kang JS, Lien D-H, Najmzadeh M, Tongay S, Wu J, Javey A (2015) MoS_2 heterojunctions by thickness modulation. Sci Rep 5:10990

Ueno K, Saiki K, Shimada T, Koma A (1990) Epitaxial growth of transition metal dichalcogenides on cleaved faces of mica. J Vac Sci Technol A 8:68–71

Utama MIB, Zhang Q, Zhang J, Yuan Y, Belarre FJ, Arbiol J, Xiong Q (2013) Recent developments and future directions in the growth of nanostructures by van der Waals epitaxy. Nanoscale 9:3570–3588

Utama MIB, Kleemann H, Zhao W, Ong CS, da Jornada FH, Qiu DY, Cai H, Li H, Kou R, Zhao S, Wang S, Watanabe K, Taniguchi T, Tongay S, Zettl A, Louie SG, Wang F (2019) A dielectric-defined lateral heterojunction in a monolayer semiconductor. Nat Electron 2:60–65

Valdivia A, Tweet DJ, Conley JF (2016) Atomic layer deposition of two dimensional MoS_2 on 150 mm substrates. J Vac Sci A 34:021515

van Donkelaar J, Yang C, Alves ADC, McCallum JC, Hougaard C, Johnson BC, Hudson FE, Dzurak AS, Morello A, Spemann D, Jamieson DN (2015) Single atom devices by ion implantation. J Phys Condens Matter 27:154204

Wachter S, Polyushkin DK, Bethge O, Mueller T (2017) A microprocessor based on a two-dimensional semiconductor. Nat Commun 8:14948

Wang L, Meric I, Huang PY, Gao Q, Gao Y, Tran H, Taniguchi T, Watanabe K, Campos LM, Muller DA, Guo J, Kim P, Hone J, Shepard KL, Dean CR (2013) One-dimensional electrical contact to a two-dimensional material. Science 342:614–617

Wang S, Wang X, Warner JH (2015) All chemical vapor deposition growth of MoS_2: h-BN vertical van der Waals heterostructures. ACS Nano 9:5246–5254

Whitener KE Jr, Sheehan PE (2014) Graphene synthesis. Diam Relat Mater 46:25–34

Ya Y, Qui J, Guo W, Qi Z, Zhang Y (2010) Electrical instability and negative differential resistance in single Sb-doped ZnO nanobelts/SiO_x/p-Si heterostructure device. Appl Phys Lett 96:093107

Zhang W, Huang Z, Zhang W, Li Y (2014) Two-dimensional semiconductors with possible high room temperature mobility. Nano Res 7:1731–1737

Zwanenburg FA, Dzurak AS, Morello A, Simmons MY, Hollenberg LCL, Klimeck G, Rogge S, Coppersmith SN, Eriksson MA (2013) Silicon quantum electronics. Rev Mod Phys 85:961–1012

Chapter 2
Atomic Electronics

Abstract Atom electronics is the final scale of electronics, where electronic devices are based on a single or few atoms. The electronic circuits have evolved from micro to nanoscale, an example being circuits based on process technologies developed for FinFETs at 10 nm node. Going bellow with scaling, we reach the atomic scale, where the first devices based on single-atom-like structures such as single dopants are reported and analyzed.

Although transistors based on a single or few dopants were recently reported, atomic scale electronics or atom electronic is at the very beginning, mainly due to many difficulties in fabrication and replication, as well as due to the working regime, which requires mainly very low temperatures. The goal of field-effect-transistors (FETs) based on atomic-like structures is to control the conductivity at the atomic level via quantum confinement, i.e., by creating and exploiting discrete energy states. The typical configuration of an atom-like FET is based on a dopant (the atom-like nanostructure) placed between source (S) and drain (D) electrodes of the transistor, its discrete energy levels being controlled by a gate (G), as shown in Fig. 2.1.

As we have pointed out in the previous chapter, single or few dopants can be introduced in the channel of a FET with advanced nanotechnology equipments such as scanning tunneling microscope (STM) or atomic force microscope (AFM), both able to manipulate single atoms on surfaces.

In the single-dopant transistor, single-electron tunneling is taken place when the dopant is placed in the channel of the FET (Moraru et al. 2011) and this is recognized in the drain current (I_D)–gate voltage (V_G) dependence at low temperatures. The dopant is acting as a quantum dot, i.e., as an atom-like structure with two or more discrete energy levels, and there is a tunneling conduction path activated by the gate voltage when Fermi level is aligned with the donor energy levels. Then, a sharp maximum in the $I_D - V_G$ characteristics is observed, as illustrated in Fig. 2.2.

In a rich doped channel (see Fig. 2.3), the $I_D - V_G$ signature of the FET consists of a series of peaks, resulting from the contributions of several individual/non-interacting quantum dots. In both cases, although the gate voltage is typically in the range 1–3 V, the drain current is very weak, of few picoampers, even when measurements are made at 10–15 K.

M. Dragoman and D. Dragoman, *Atomic-Scale Electronics Beyond CMOS*,
https://doi.org/10.1007/978-3-030-60563-6_2

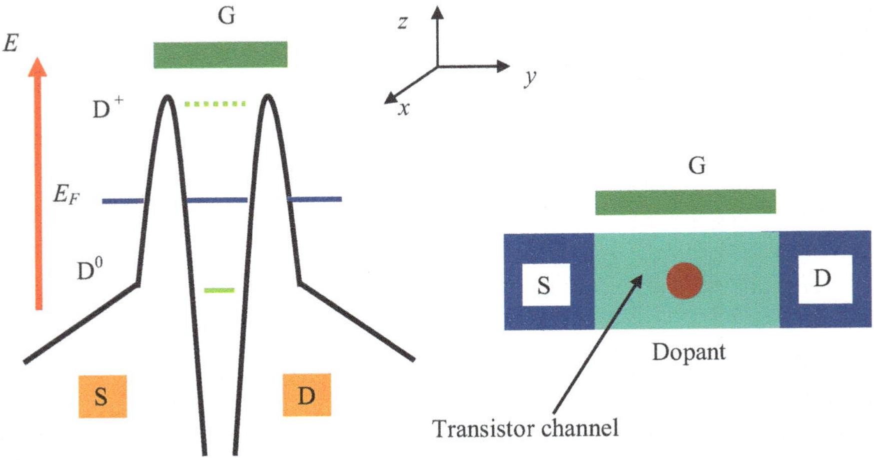

Fig. 2.1 Potential energy (left) and schematic representation of a dopant (atom) introduced in a FET channel

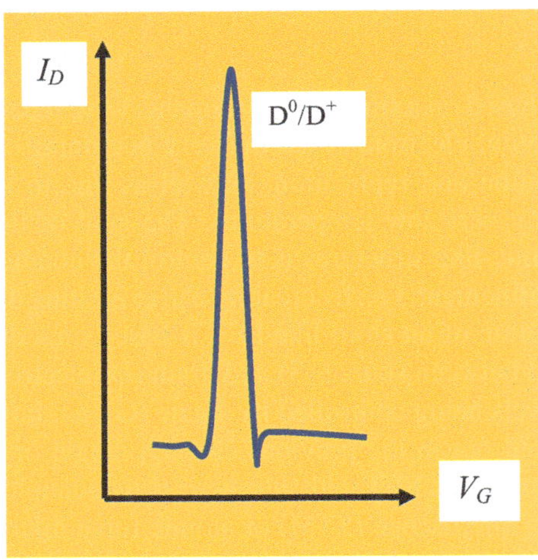

Fig. 2.2 I_D–V_G signature of a single dopant in a FET at low temperature

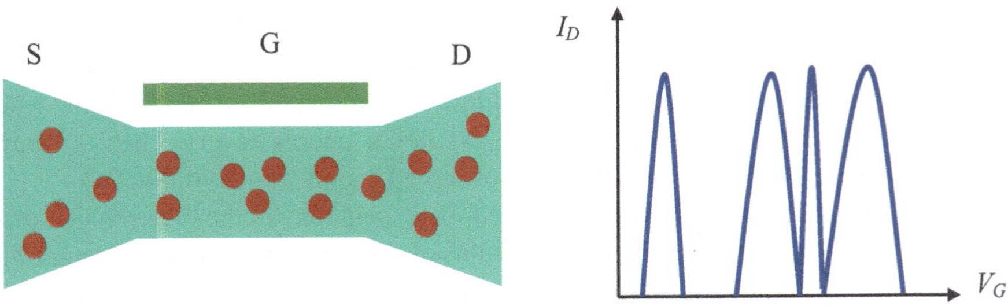

Fig. 2.3 FET channel with multiple dopants and its I_D–V_G signature

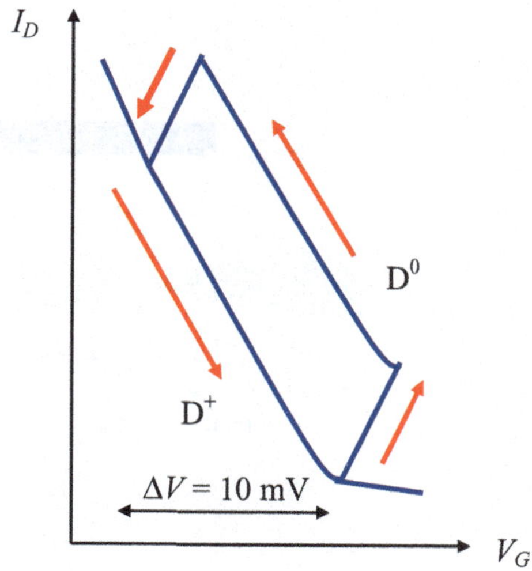

Fig. 2.4 I_D–V_G dependence of the coupled dopants (adapted from Hamid et al. 2010)

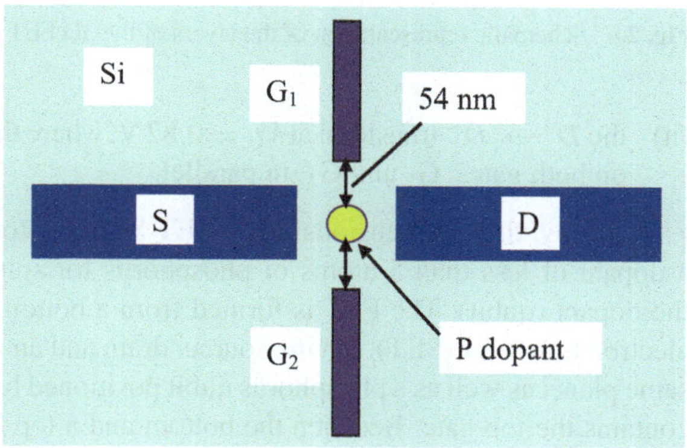

Fig. 2.5 The single atom transistor on Si

More interesting effects occur when two donors are coupled. In this situation, by sweeping the gate voltage, charging and discharging effects are observed for the energy levels corresponding to the D^0 and D^+ states, such that a hysteretic behavior appears, which could be used for logical operations or memories (see Fig. 2.4) (Hamid et al. 2010). Again, the currents are weak, of the order of 0.1 ps.

Starting from these initial remarkable results, which can be followed in more depth in the cited references above and in the review (Zwanenburg et al. 2013; Fuechsle et al. 2017), a single atom transistor on Si was recently reported, which operates at milliKelvin temperatures (see Fig. 2.5). In this case, using STM, a single phosphorus atom dopant was positioned at a precise place in the channel of a FET transistor starting from a precursor desorption, followed by its dissociation and rejection of a Si ad atom and incorporation of a P atom. Measuring the $I_D(V_G)$ dependences, transitions between dopant charge states, the following facts were evidenced:

(i) the $D^+ \to D^0$ transition at $V_G = 0.45$ V, and

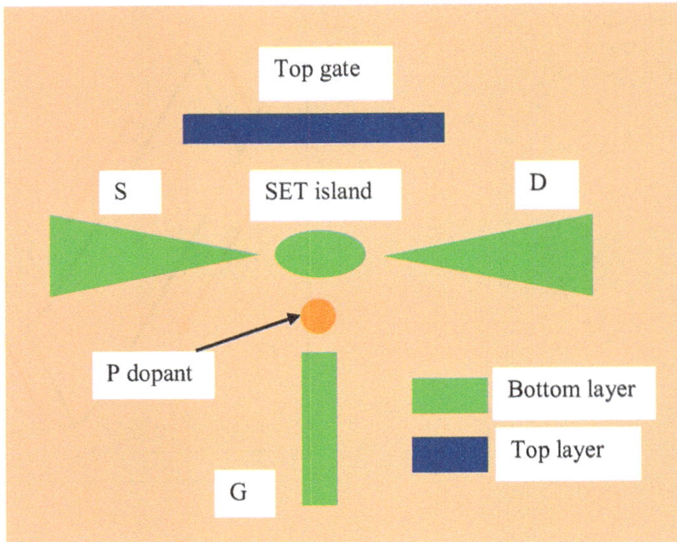

Fig. 2.6 Schematic representation of the layers of the 3D FET

(ii) the $D^0 \rightarrow D^-$ transition at $V_G = 0.82$ V, where the gate voltage V_G was applied on both gates, G_1 and G_2, in parallel.

Recently, three dimensional (3D) FET Si transistors were fabricated containing a dopant of less than 5 atoms of phosphorus for spin readout of the spin state in the dopant (qubit). The FET is formed from a bottom layer, which contains single electron transistor (SET), having source, drain and an in-plane gate electrodes in the same plane, as well as a phosphorus qubit positioned by STM. The second, top layer, contains the top gate. Between the bottom and a top layer is an epitaxial grown Si layer with a thickness of 40 nm (Koch et al. 2019). The 3D FET atomic transistor is displayed schematically in Fig. 2.6. In principle, when the spin is down there is no drain current in the SET, while when the spin is up, the drain current is high when a magnetic field >1 T is applied on the FET.

All the above FET configurations are working at very low temperatures, because the dopants located in the channel of the transistor have shallow electronic states, i.e., a potential well of only 100 meV. In contrast, if the dopant is placed inside a spatially-confined insulator (a point-contact neck), a deep potential of 2 eV is formed at room temperature, with multiple discrete electron states. Therefore, it is possible to have single-dopant transistors at room temperature based on 10 nm $Si/SiO_2/Si$ point-contact tunnel junctions, which are fabricated by scanning probe lithography and geometric oxidation (Durrani et al. 2018) (see Fig. 2.7). The FET has a back-gate configuration. The discrete levels induced by dopants were estimated by simulations and checked by measurements of the transistor's parameters.

Another atomic-scale device is the atomic switch, which is a two-terminal metal-oxide-metal structure able to connect or disconnect an atomic-like wire filament

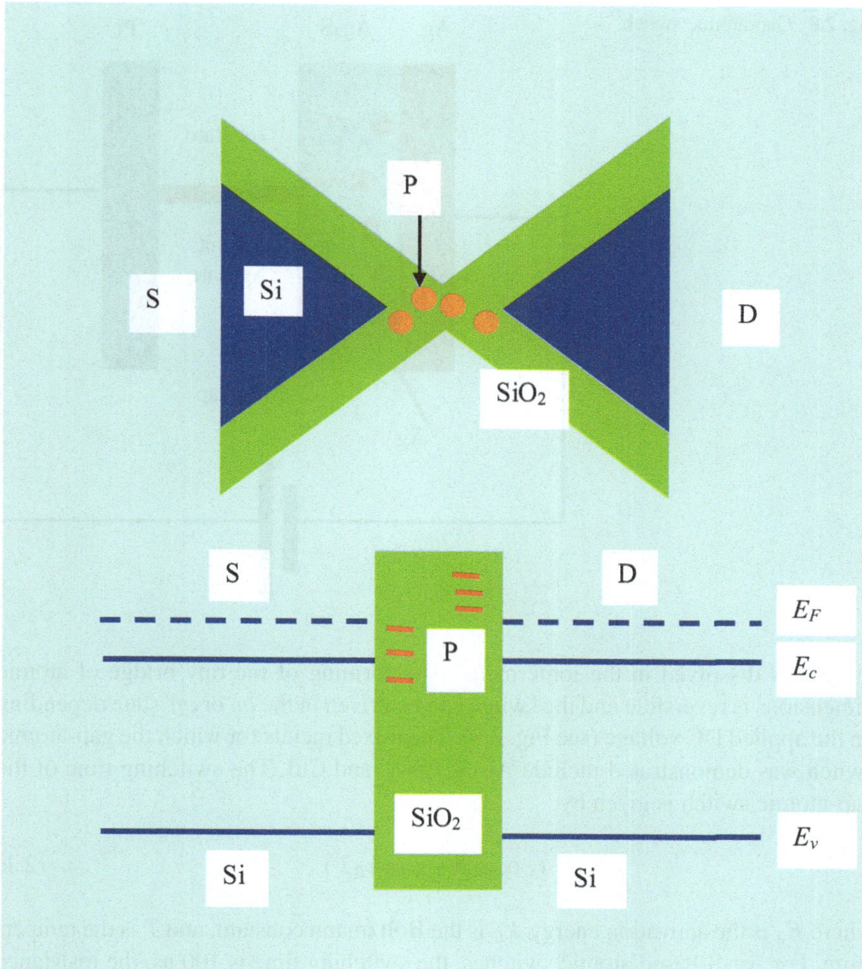

Fig. 2.7 The single-dopant FET working at room temperature (top) and its energy diagram (bottom). The Si regions are degenerately doped (they act as S and D electrodes). In the neck region, two sets of discrete states, for two P dopants, are shown

(Aono and Hasegawa 2010). These are electrochemical switches, which, based on diffusions of ions and their oxidation, are creating or destroying a metallic path between two electrodes.

There are several types of atomic switches. The first type, termed gap-atomic switch, consists of two metallic electrodes (Ag and Pt) separated by an electronic mixed conductor containing cations—an ionic conductive material (Ag_2S), and a nanogap. When a positive voltage is applied between the two electrodes, the metal cations Ag^+ are flowing in the nanogap and form a wire-like bridge, which puts the switch in the *on* state. Otherwise it is in the *off* state, in which there is ionization

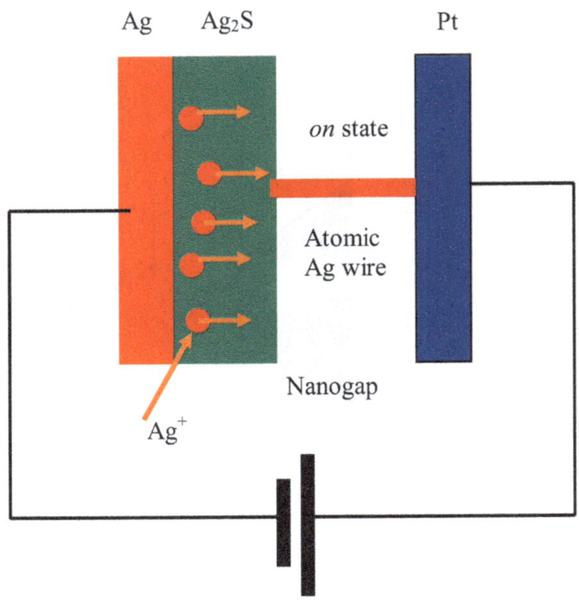

Fig. 2.8 Gap-atomic switch

of cations dissolved in the ionic metal. The forming of the tiny bridge of atomic dimensions is reversible and the switch can be driven in the *on* or *off* state depending on the applied DC voltage (see Fig. 2.8). The mixed metals for which the gap-atomic switch was demonstrated include Ag_2S, Cu_2S, and CuI. The switching time of the gap-atomic switch is given by

$$t_S \propto \exp(-E_a/k_B T) \tag{2.1}$$

where E_a is the activation energy, k_B is the Boltzmann constant, and T is the temperature. For Ag_2S-based atomic switches, the switching time is 100 ns, the resistance in the *on* state ranges between 100–1000 Ω, while the *off*-state resistance is typically larger than 1 MΩ and can reach even 1 GΩ.

The gapless atomic switch is the second type of atomic switches, in which the ionic path is created in an ionic conductive material (Ta_2O_5 in this example). In this case, the counter electrode (Pt) is in contact with the ionic conductive material, leaving no gap between them and trapping the diffused metal cations (Cu^+). Thus, the metal atoms are precipitating towards the counter electrode, forming an atomic-like metal filament, which puts the switch in the *on* state (see Fig. 2.9). When the polarization is reversed, the dissolution of atomic-like filament is taking place, and the switch is in the *off* state.

Both types of atomic switches are non-volatile, and thus are native memories. The gapless atomic switch was demonstrated with a series of materials, the most remarkable results being obtained in HfO_2, SiO_2 and TiO_2, which are CMOS-compatible oxides.

Fig. 2.9 The gapless atomic switch

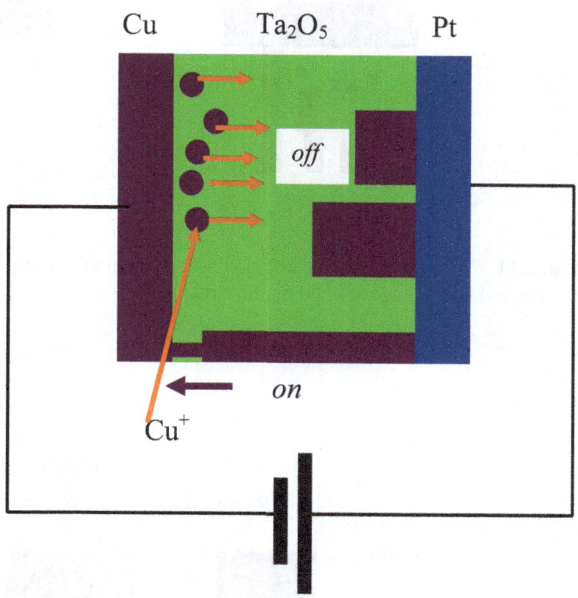

Fig. 2.10 Atomic switch based on Cu cations and HfO_2

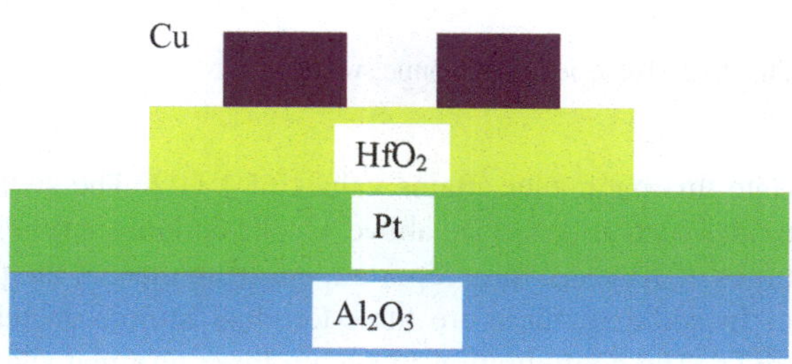

In the case of HfO_2, the atomic switch consists of the $Cu/HfO_2/Pt$ structure, having an *on*/*off* ratio of 10^3 to 10^4 (see Fig. 2.10). Cu cations are forming filamentary paths through HfO_2 up to the Pt electrodes, and are annihilated when the applied DC voltage switches polarity (Haemori et al. 2009). In terms of resistive switches, the *on* state is termed as LRS (low resistance state) and the *off* state as HRS, i.e., high resistance state. Copper is chosen as a material in such switches due to its mobile ions, which show high diffusion rates, and thus have the facility to form conductive filaments (see Fig. 2.11).

In a similar manner, $Ir/SiO_2/Cu$ could be used as atomic switch. Again, Cu cations are diffused through SiO_2 in the presence of an applied positive voltage, and form a filament, which is destroyed by reversing the polarity of the applied voltage. The write current is 10 pA (Schindler et al. 2008). A forming process is often necessary in order to initiate the physical processes in gapless atomic switches. This process implies the application of high voltages on the device until the diffusion process is initiated. The LRS state has a resistance of around 100–1000 Ω, while in the HRS

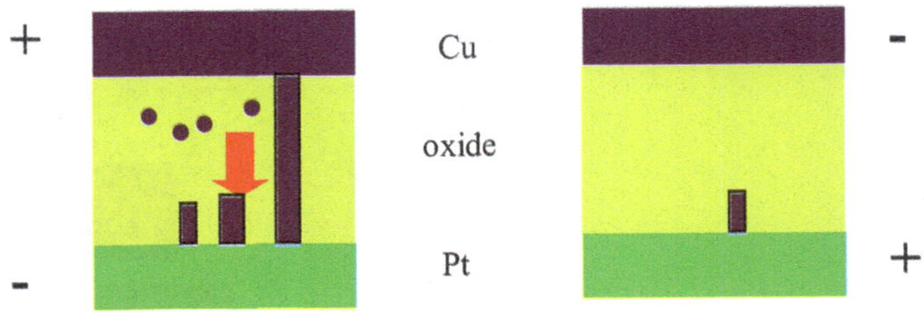

Fig. 2.11 The atomic switch Cu filament formation and destruction depending on the polarity of the applied DC voltage applied on the structure in Fig. 2.10

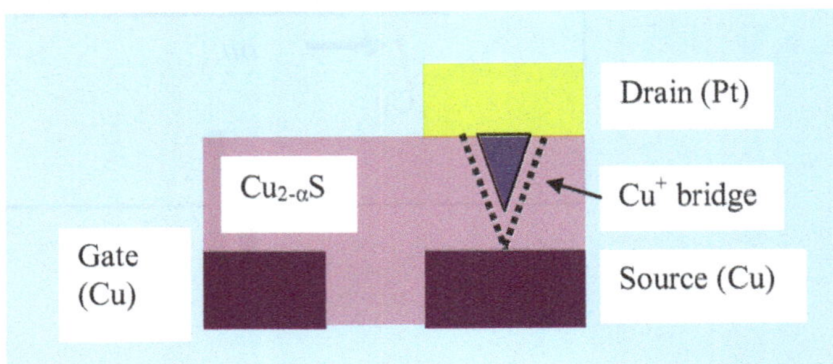

Fig. 2.12 Three-terminal atomic switch

state this parameter attains values of 1 GΩ. The switching time is 5 ns for metal oxides such as described above. Cyclic endurance times of 10^{11} to 10^{12} and retention times of 10 years were reported, ensuring thus all the properties of a good memory.

In addition, there are three-terminal atomic switches, which are analogous to FETs. The configuration of such a three-terminal switch is presented in Fig. 2.12. The gate in atomic switches has the role to control the formation and dissolution of conductive filaments (Cu^+ filaments in this case). The separation between the signal line and control line makes these three-terminal devices suitable candidates for logic circuits at atomic scale. The metal filament is formed when a positive voltage is applied to the gate and is destroyed when the polarity of the gate voltage is changed. When working with solid electrolytes the switching time is less than 10 s, while the *on*/*off* ratio is 10^5 (Banno et al. 2006).

The atomic switch concept was developed further to implement memristors, which are resistances with memory. The memristor physical properties, which will be explained below, are based on the movement of oxygen ionic carriers through oxygen vacancies at high electric fields in various dielectrics. A memristor plays a similar role to the inductance (L) and capacitance (C). The memristor is a nonlinear circuit element, which has a unique signature in the current-voltage dependence. Namely, this dependence shows a pinched hysteretic behavior of the current when the voltage is double-swept from negative to positive voltages, as illustrated in Fig. 2.13 (Chua 2014).

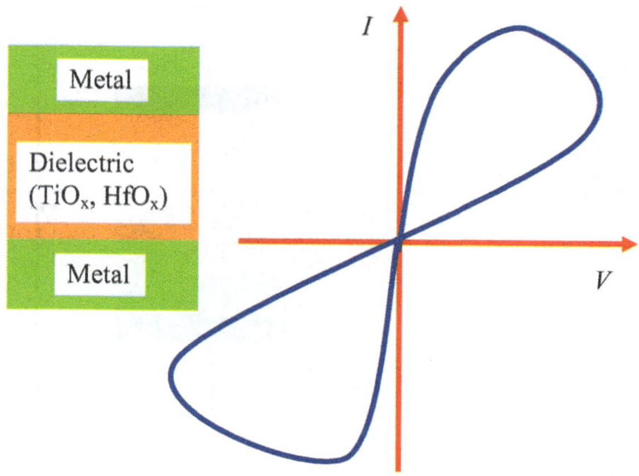

Fig. 2.13 Current-voltage dependence of a memristor

The memristor, which is a non-volatile device as any atomic switches, has an electric resistance that depends strongly on the history of the applied voltage. Memristors have numerous applications in reconfigurable logic circuits (Xia 2014) or neuromorphic systems (Jo et al. 2010), since a memristor is an electrical analogue of the synapse.

The time dependence of the current-voltage characteristics of memristors can be expressed as (Jo et al. 2010):

$$i(t) = \frac{v(t)}{R_{on}\gamma(t) + R_{off}[1 - \gamma(t)]} \tag{2.2}$$

where $\gamma(t)$ is a continuous function of time, with values in the [0,1] domain. The function $\gamma(t)$ reaches its maximum and minimum values, of 1 and 0, when $R = R_{on}$ (LRS) and, respectively, $R = R_{off}$ (HRS).

The first memristor (Strukov et al. 2008) was based on a vertical structure formed from two regions of TiO_2, one of them being a TiO_{2-x} region with oxygen deficiency of 2.5%, sandwiched between two Pt electrodes. Since the oxygen vacancies are acting as positive charges, the entire structure can be seen as a heterostructure metal/doped (conductive) material/undoped (insulator) material/metal, as shown in Fig. 2.14.

The general equations describing a current-controlled memristor are written as (Strukov et al. 2008):

$$v = R(w)i \tag{2.3a}$$

$$\mathrm{d}w/\mathrm{d}t = i \tag{2.3b}$$

In the case of oxide memristors based on oxygen vacancies dopants having the mobility μ_V we have

Fig. 2.14 Memristor based on TiO_2

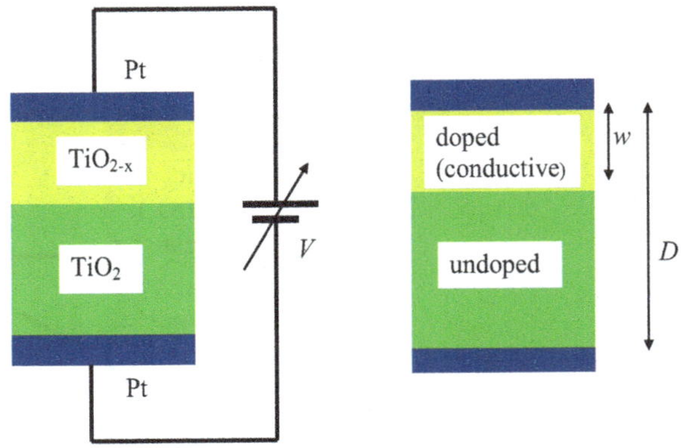

$$v(t) = \left(R_{on}\frac{w(t)}{D} + R_{off}\left(\frac{1 - w(t)}{D}\right)\right)i(t) \tag{2.4}$$

$$\mathrm{d}w(t)/\mathrm{d}t = \mu_V(R_{on}/D)i(t) \tag{2.5}$$

and thus

$$w(t) = \mu_v(R_{on}/D)q(t) \tag{2.6}$$

where D is the thickness of the semiconductor layer placed between the metal electrodes and q is the charge. Finally introducing (2.6) in (2.4), and considering that $R_{on} \ll R_{off}$ we have the memristance $M(q)$ as

$$M(q) \cong R_{off}\left[1 - \mu_V(R_{on}/D^2)q(t)\right] \tag{2.7}$$

The equations above describe the dynamics of the memristor in the doped/undoped states. When a positive voltage is applied to the memristor, the doped (conductive layer) is increased, whereas when a negative voltage is applied the conductive ionic filaments are disabled. The positive polarization is attributed to R_{on} and the negative polarization to R_{off}. At sweeping several times the voltage, the memristor behaves as shown in Fig. 2.15.

Beyond the applications of memristors in memories and neuromorphic computation, which will constitute a separate chapter of this book, we remark that, recently, microwave switches were reported using the concept of atomic switches based on ions, discussed earlier, and memristors (Pi et al. 2015). Thus, processes at atomic scale are not used only in miniaturized devices at the nanoscale, but in electronic devices having dimensions much larger than the tiny atomic switches integrated in them. In the case of microwave switches, the Ag active metal, which forms conductive filaments, is patterned with a gap of 35 nm on top of a much wider gap in a gold electrode, as illustrated in Fig. 2.16. The Ag electrode is 70 nm thick and the

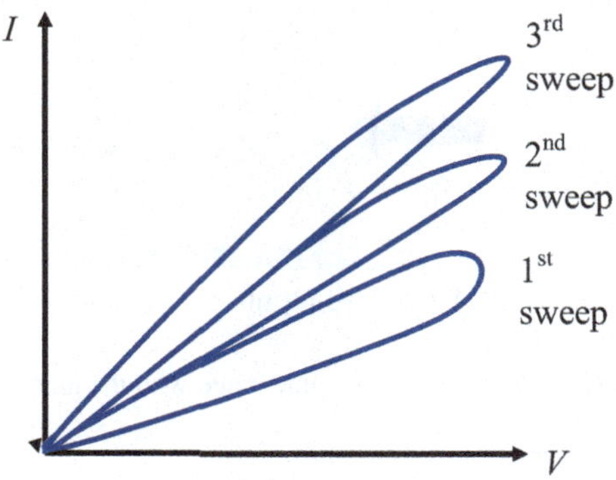

Fig. 2.15 The memristor imprint

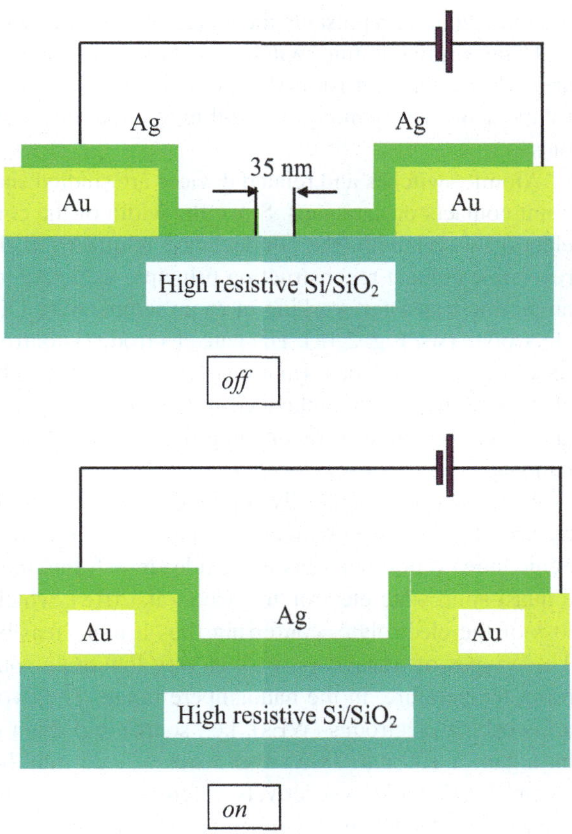

Fig. 2.16 Microwave atomic switch

thickness of gold electrodes is 140 nm. Each electrode is connected to Cu/Al electrodes with thicknesses >1 μm. The atomic switch is in the *on* state at 3 V (when the conductive filaments bridge the gap) and the *on*/*off* ratio is 10^{12}. The Au electrode on which the Ag metal is patterned is part of a coplanar line, i.e., is placed between

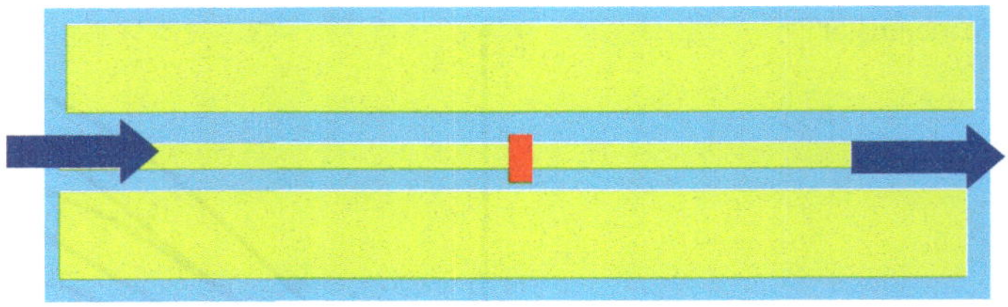

Fig. 2.17 Coplanar line integrated with the memristor atomic switch

two outer Au electrodes that form a coplanar line at 40 GHz, with 0.3 dB insertion loss and 30 dB isolation, as shown in the configuration depicted in Fig. 2.17, where the red rectangle represents the memristor atomic switch.

Using similar atomic switches as above, we have designed and simulated antenna arrays based on memristors (Dragoman et al. 2017). The antenna area beam is steered by the action of memristors, exhibiting a maximum steering angle of ±28° at 2 V bias.

Atomic switches and related devices are studied since many years using metallic point contacts or nanogaps. Since the width of the contacts is comparable with the electron wavelength, the conductance is quantized in steps of $G = 2e^2/h$. The reversible contact reconstruction principle using Ag, explained above, was used in an atomic transistor working at room temperature (Xie et al. 2008) employing an electrolyte (see Fig. 2.18). The gate electrode G controls the potential distribution in the electrolyte, and the formation of Ag filaments that bridge the gap between the Au electrodes, which are embedded in the electrolyte. Thus, it is possible to precisely and reversibly control the opening (*off* state) and the closing (*on* state) of the tiny nanogap.

The electrolyte is usually liquid, e.g., a solution of $AgNO_3 + HNO_3$, but solid electrolytes are also possible (Xie et al. 2018). For instance, the gelation with pyrogenic silica (colloidal silicon dioxide) transforms the liquid silver electrolyte into a quasi-solid-state electrolyte (Xie et al. 2018), which allows the reduction of the loss of the electrolyte, conferring thus a more feasible solution for such devices. These types of transistors work at very low drain-source voltages (10–20 mV), at room temperature, in the nanoampere range. The two gold electrodes in Fig. 2.18 are working electrodes (WEs), i.e., source and drain electrodes, and are separated by a gap of 100 nm. Two silver wires of 0.25 mm diameter were used as counter electrode (CE) and, respectively, reference electrode (RE), their combination being termed as gate. The gate voltage is applied between CE and ground (for details on the circuitry, see Fig. 2.19)

Although the drain-source voltage is only −12.9 mV, and the conductance in time is varying between 0 and 2–5 G_0, with $G_0 = 2e^2/h$, depending on the gate voltage, which does not exceed −40 mV, this type of device is still in its infancy despite few decades of research. For a review of the research efforts invested to fabricate this

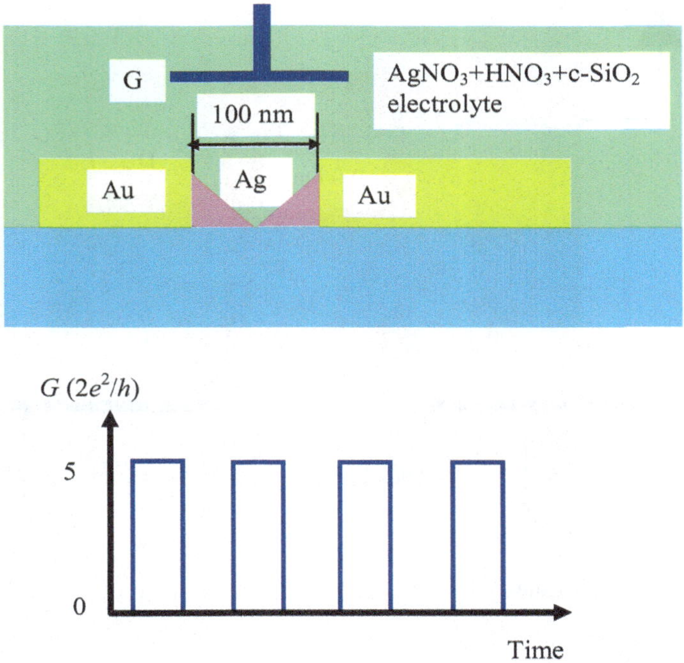

Fig. 2.18 Metallic point contact (nanogap) atomic switch and its reversible control of conductance

device and other similar devices based on other metals, such as Cu or Pb, see (Xie et al. 2018) and the references therein.

The idea to insert nanogaps between electrodes for resistive switching has inspired other studies. In principle, nanogaps are two metallic electrodes with nanometer distances between them, displaying a reversible resistance switch as a function of the voltage applied between them (see Fig. 2.20). This resistive switching is the result of the migration of atoms between the nanogap terminations, which is changing the gap width, and thus the tunneling transport between the two metallic electrodes at a certain threshold voltage, depending on the material used.

The study of metallic gap resistive switching involved mainly two types of electrodes: Au and Pt (Suga et al. 2012). The metals were deposited on a 300 nm Si over which SiO_2 was thermally grown. Both Au and Pt electrodes where used to fabricate nanogaps with a width of 10 nm (see Fig. 2.21). The Au electrodes nanogap was measured at room temperature in an environment containing water vapors at different pressures, ranging from 10 Pa to 1 kPa, and it was observed that a hysteresis in the current-voltage dependence is occurring only at 10 Pa. In contrast, Pt electrodes, which are more stable, show a hysteresis in the current–voltage dependence in ambient conditions (see Fig. 2.22). The model of the current (I)–voltage (V) dependence is based on tunneling emission of electrons, a phenomenological equation extracted from fitting curves being (Suga et al. 2010).

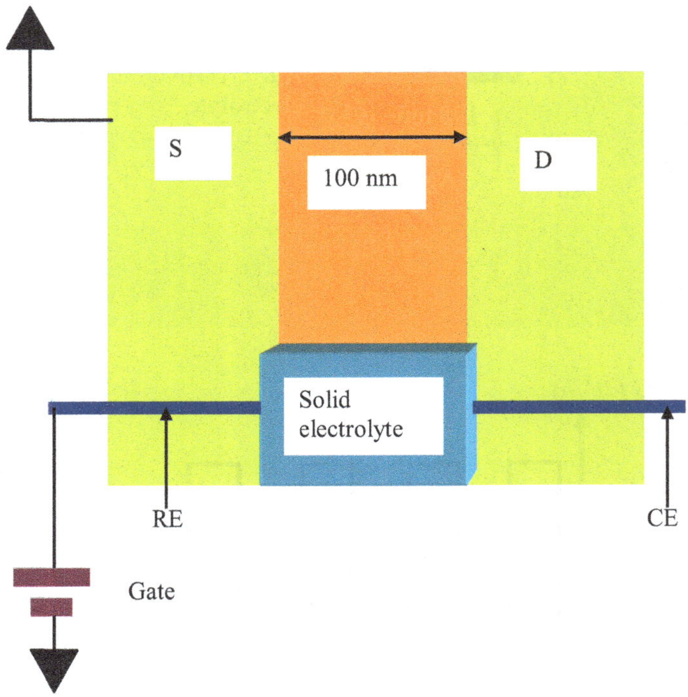

Fig. 2.19 The solid-state-electrolyte single-atom transistor

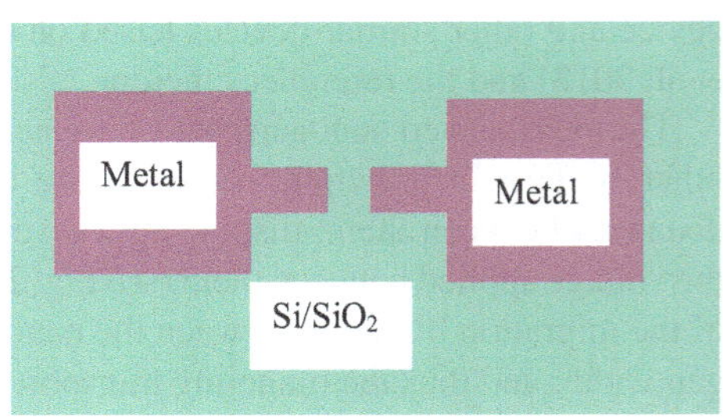

Fig. 2.20 Metallic nanogap resistive switching

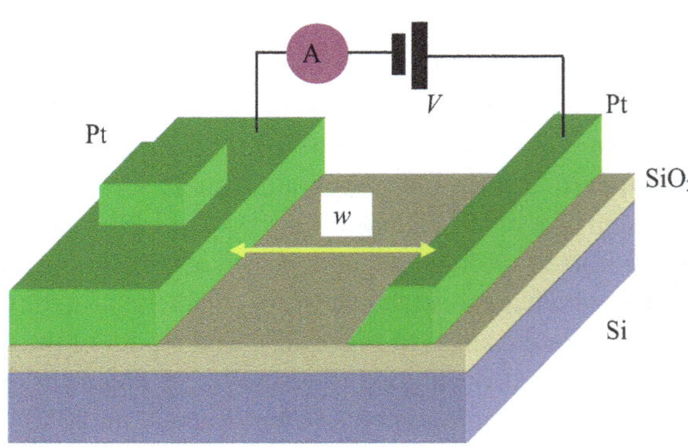

Fig. 2.21 Resistive switch working at room temperature in ambient conditions (after Suga et al. 2012)

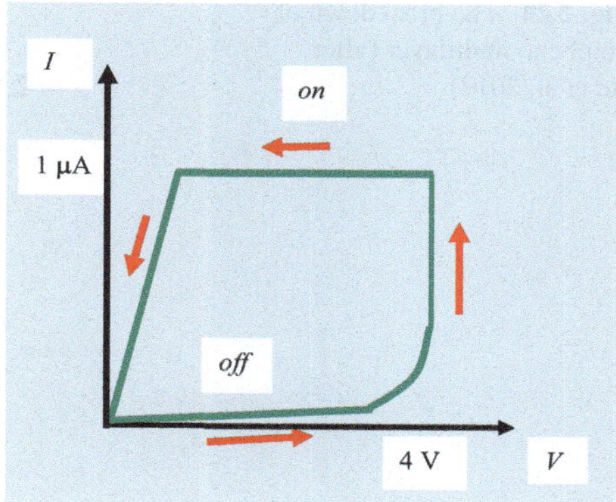

Fig. 2.22 The resistive switch of Pt nanogap at room temperature and ambient conditions (after Suga et al. 2012) with the current limited at 1 μA

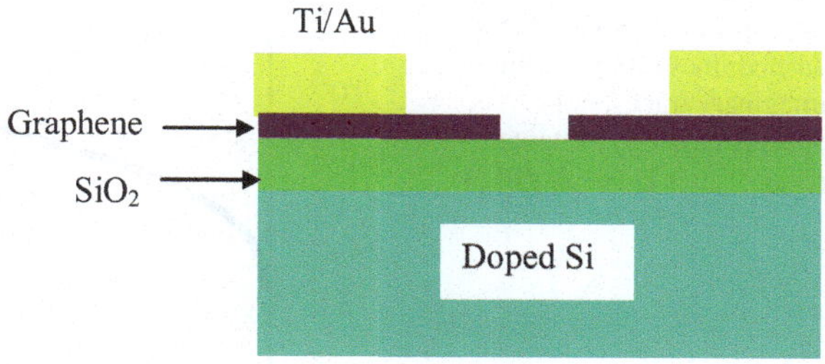

Fig. 2.23 Graphene nanogap resistive memory

$$I = (const \times k_1 A/w^2)[x^2 \exp(-k_2 wx) - y^2 \exp(-k_2 wy)] \tag{2.8}$$

where $x = (\phi - V/2)$, $y = (\phi + V/2)$, k_1, k_2 are constants, and w, A and ϕ are the gap width, tunneling area and the barrier height, respectively.

Graphene plays an important role in nanogap memories (He et al. 2012). For this application, multilayer graphene is transferred on a Si/SiO_2 substrate and two metallic electrodes of Ti/Au (2 nm/30 nm) are deposited over it. The graphene is patterned in the shape of a ribbon with the width of 1 μm and length of 450 nm given by the distance between the metallic contacts (see Fig. 2.23).

The multilayer graphene has a thickness of 2.3 nm, i.e., it has a thickness equivalent with 4 atomic layers. There is a forming process to create a nanogap of 20–40 nm via excitation with a DC voltage. Figure 2.24 shows that, by increasing the voltage at 5.5 V, the graphene breaks and the current reaches almost zero values.

After the breakdown of graphene multilayer has taken place, graphene can be used as a resistive switch and memory. However, it was observed that the oxide below graphene is partially destroyed during the heating process, which is the result of the forming process. During heating the temperature is estimating to rise at 1700 °C and Si nanocrystals filaments are formed with a length of 13 nm and a tunneling distance of 0.7 nm, determined via fitting the current-voltage dependence in the *on* state.

The current-voltage of the graphene nanogap is displayed in Fig. 2.25. The two states specific for resistive switching, LRS and HRS, can be observed. The transition

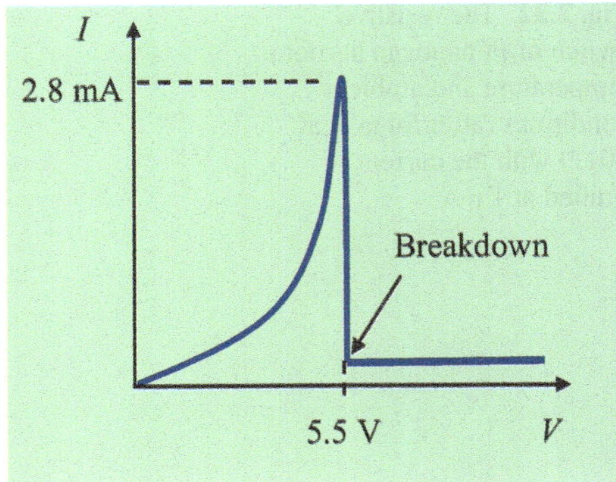

Fig. 2.24 The breakdown of graphene multilayer (after He et al. 2012)

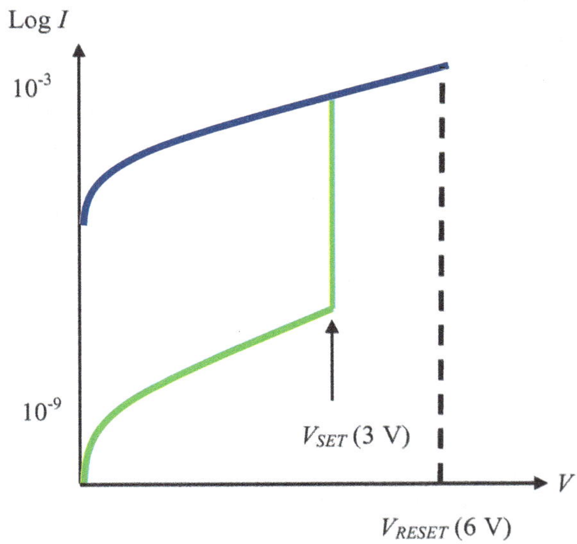

Fig. 2.25 h and LRS states of graphene nanogap; the dashed lines represent the imaginary traces, not appearing in experimental results (after He et al. 2012)

from LRS to HRS takes place at a certain set voltage, denoted by V_{set}, and the transition from HRS to LRS occurs at the reset voltage, V_{reset}. Both states are nonvolatile, and the tunneling current density is given by

$$J = I/A = V^2 \exp(-1/V) \tag{2.9}$$

which describes the Fowler-Nordheim tunneling process via Si nanocrystals filaments formed in the graphene nanogap, part of them being reoxidized during reset process.

The type of switching behavior is another important issue. Unipolar and bipolar switching refer to the creation of the two states via set and reset voltages, which have the same polarity or opposite polarities, respectively. In our case, the two states are

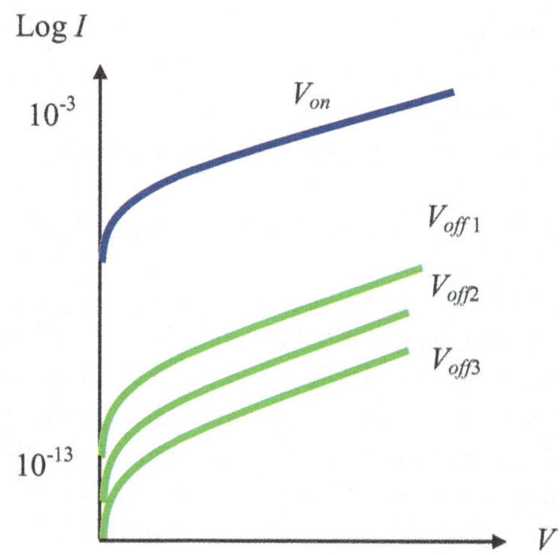

Fig. 2.26 Multilevel memory

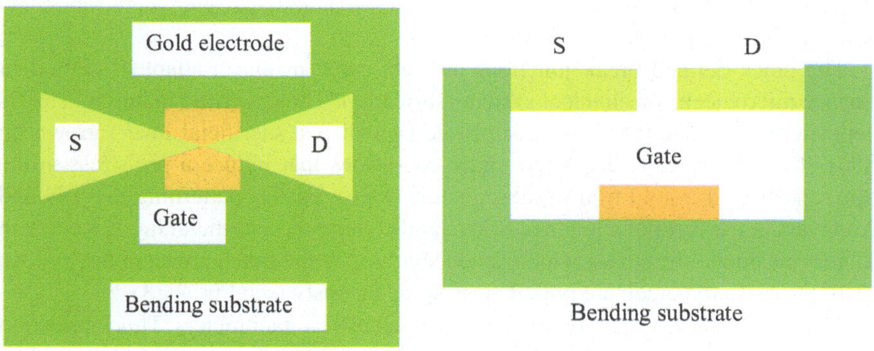

Fig. 2.27 Top view (left) and side view (right) of the atomic NEMS switch

created via a non-polar switching mechanism, i.e., they are produced by any combination of set and reset voltages, i.e., positive-negative, positive-negative, negative positive, or negative-negative voltages, which shows that the physical mechanism of switching is related to the heating process and not to the electric field. Moreover, multi-level memory switching can be obtained by lowering the reset voltage, as shown in Fig. 2.26.

There is an interesting version of nanogap switches based on the nanoelectromechanical (NEMS) concept working in vacuum (Martin et al. 2009). Basically, two gold electrodes, which are very narrow at their ends, are fabricated on polyimide deposited on a phosphor bronze wafer. Using the sacrificial layer method, the two gold electrodes are forming a bridge suspended over a buried metallic gate separated by vacuum (see Fig. 2.27). Measurements of this device are made at 6 K in vacuum, and the gap between the metallic electrodes is open or closed by bending the substrate with a screw and a servomotor. Thus, the conductance of the device is modulated from $2e^2/h$ (*on* state) up to much lower values obtained in the tunneling regime (*off* state), depending on the gate voltage.

The modulation of the conductance is described by the formula:

$$G(L + \Delta L) = G(L)\exp[-(8m\phi)\Delta L/\hbar] \tag{2.10}$$

where L is the distance between electrodes and ΔL is the change of this distance, ϕ is the work function of gold, which is 5.4 eV, and m is the electron mass.

The nanogaps are fabricated by different methods using e-beam lithography, AFM or STM. An updated review regarding the fabrication methods is found in Yang et al. (2019). Mechanical breaking is a typical method to fabricate nanogaps, as described in the example above. However, none of the already mentioned methods are suitable for large scale fabrication, where millions of devices are fabricated in a single step. This issue is typical for many devices at atom-scale. Nonetheless, in the case of the metallic nanogaps required in many single-mole devices, a massive parallel fabrication of crack-defined gold break junctions was very recently demonstrated (Dubois et al. 2018). In this respect, 7 millions junctions per cm^2 were fabricated with sub-3 nm gap with an yield of 7%.

The crack-defined break junctions are fabricated using an adapted method of surface microelectromechanical system fabrication. Thus, a brittle material (TiN) is deposited on an amorphous a-Si substrate (which is a sacrificial layer) grown on SiO_2. The TiN layer is deposited under conditions that induce a pre-stress in the film, and over it a gold film is subsequently deposited. By patterning the TiN and Au layers in the form of a notched bridge and etching the sacrificial layer, a crack in TiN forms due to the stress, which is concentrated in the notch area and the two Au sections above the crack are pulled apart spontaneously (see Fig. 2.28).

Thus, two cantilevers are produced, separated by few nanometers. This fabrication method was applied on a 100 mm wafer, and 780,000 bridge structures were fabricated in a single step, 95% of them being simultaneously released using isotropic plasma etching.

Graphene is the material of choice at atomic scale due to its many unique physical properties and phenomena. Among these, graphene nanoribbons and graphene constrictions are configurations of interest, Klein tunneling and atom collapse being some of the specific physical phenomena that are encountered in graphene.

Graphene nanoribbons (GNRs) are graphene stripes having widths W not exceeding few tens of nanometers. If the x axis is chosen as the longitudinal axis of the GNR, then the wave vector is quantized in the transverse direction, y, along which it takes discrete values $k_y = n\pi/W$, where n is an integer number. Thus, the dispersion relation of graphene is changed dramatically from its linear form, and a bandgap appears in GNR as a result of the symmetry breaking. According to Fang et al. (2007), the dispersion relation becomes:

$$E(n, k_x) = s\hbar v_F\sqrt{k_x^2 + (n\pi/W)^2} \tag{2.11}$$

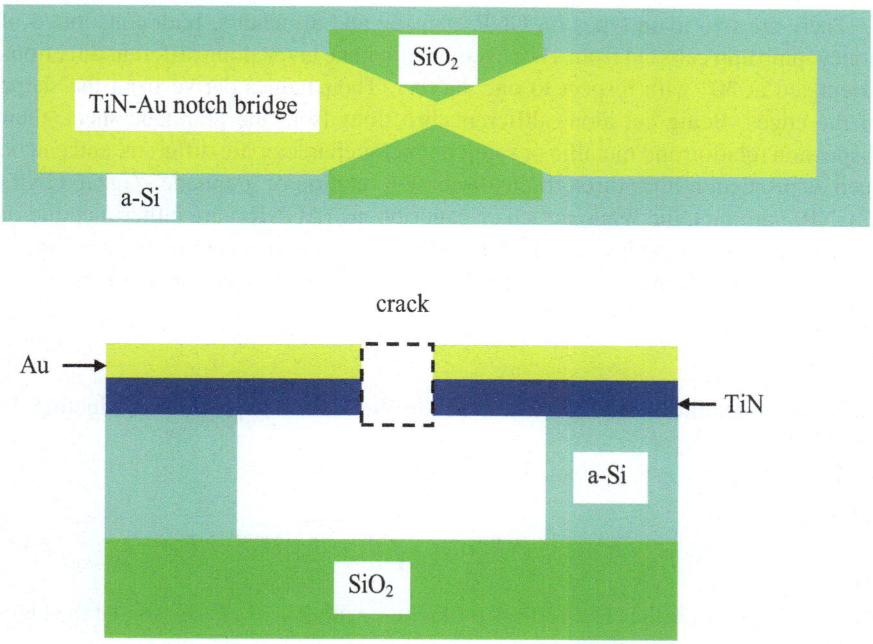

Fig. 2.28 Crack-defined break junction for batch fabrication: before breaking, top view (top), after breaking, side view (bottom)

Table 2.1 The bandgap of GNR

Width (nm)	E_g (meV)
1	1000
5	500
10	100
60	4
90	2

The symmetry breaking has as main effect the breaking of the continuous conduction and valence bands of graphene into 1D subbands, where $s = 1$ refer to the conduction band and $s = -1$ to the valence band. The bandgap is inversely proportional to the graphene width W, such that narrower graphene ribbons show larger bandgaps. We have:

$$E_g = E_{s=+1}(1,0) - E_{s=-1}(1,0) = 2\pi\hbar v_F/W \tag{2.12}$$

so that the bandgap depends only on the width of the GNR. We present below, in Table 2.1, some experimental values of the bandgap at different widths (Han et al. 2007; Geng et al. 2017). In principle, beyond 100 nm the bandgap is vanishing.

There are two main types on GNRs: zigzag and armchairs, both constituted of honeycomb unit cells cut from a honeycomb graphene layer along different directions orientated at 30° with respect to one another. Their names derive from the shape of the edges. Being cut along different directions from the graphene sheet, their dispersion relation and metallic or semiconductor character are different, and can be seen as projection lines through the dispersion relation of graphene. Zigzag GNRs (ZGNRs) are metallic while armchair nanoribbons (AGNRs) are either metallic or semiconducting, in the last case the bandgap depending inversely with the width, as discussed above. If the width is formed from N carbon dimmers lines, when

$$N = 3M - 1 \tag{2.13}$$

the GNR is metallic while otherwise GNR displays a bandgap/is semiconducting. In (2.13) M is an integer.

The DOS of the n-th subband is given by

$$\rho_{\mathrm{GNR}}(n, E) = (4/\pi\hbar v_F)(E/\sqrt{E^2 - E_n^2})\mathrm{H}(E - E_n) \tag{2.14}$$

where $\mathrm{H}(E - E_n)$ is the Heaviside step function, with $E_n = nE_g/2$, so that the DOS of the GNR can be calculated as

$$\rho_{\mathrm{GNR}}(E) = \sum_n \rho_{\mathrm{GNR}}(n, E) \tag{2.15}$$

Another parameter of interest is the carrier density. The carrier density of the GNR can be tuned by an applied gate voltage, as in the case of graphene, and is expressed as

$$n_{\mathrm{GNR}} = (4/\pi\hbar v_F)\sum_n (E_F^2 - E_n^2)^{1/2}\mathrm{H}(E_F - E_n), \tag{2.16}$$

where $eV_g = E_F$.

However, the mobility in GNR is decreasing as the graphene width is decreasing, so a compromise must be found in practice between a rather high mobility and a suitable bandgap value. For example, at $W = 15$ nm (which could be obtained via e-beam lithography followed by oxygen plasma etching), the mobility is 1500 cm^2/V s for GNR on SiO_2, a value which is about for times larger than in Si MOS transistors (see Table 1 in Geng et al. 2017).

The GNRs are fabricated by different methods and good reviews in this respect are Geng et al. (2017) and Celis et al. (2016). We summarize them here briefly to know what can be expected for device performances (see Fig. 2.29).

As can be seen from Fig. 2.29, there are many methods for GNR fabrication, some of them being not even listed in this figure, especially top-down approaches, which can also provide very good results. The simulations made in Geng et al. (2017) have shown that the GNR has a lower sub-threshold than Si MOS transistors (60 meV/dec).

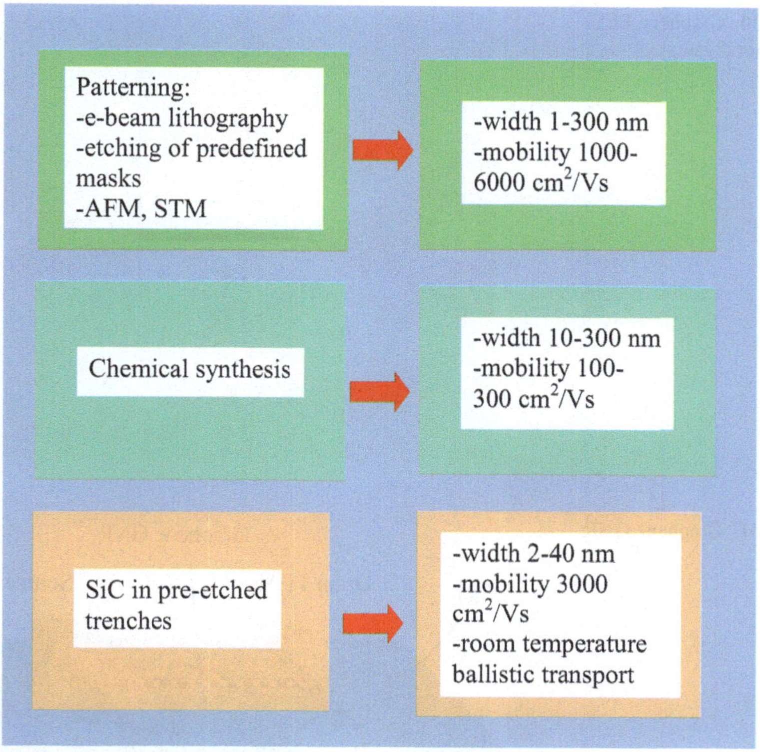

Fig. 2.29 Main methods of fabrication of GNR and performances

In RF technology, the performances of GNR FETs are comparable to those of InP HEMTs, the best transistors to date working in the THz regime at room temperature (Deal et al. 2011).

For nanoelectronic applications, top gates are difficult to fabricate for GNRs with very small widths, case in which the isolation oxide could also destroy the GNR channel when it is grown. So, an isolation-free gate solution is the lateral gate displayed in Fig. 2.30.

In this type of GNR FETs, the side gate capacitance is represented by the fringing capacitance, given by (Hähnlein et al. 2012):

$$C_f = (\varepsilon_r + 1)\varepsilon_0 \sqrt{K(1-k^2)}/K(k) \tag{2.17}$$

where ε_r is the dielectric constant of the substrate, $K(k)$ is the elliptic integral of the first kind with argument $k = [d_{\text{G-ch}}/(d_{\text{G-ch}} + W)]^{1/2}$, where W is the GNR width and $d_{\text{G-ch}}$ the distance between the gate and the channel. The fringing capacitance per unit area is then $C'_f = C_f/(W \times L)$, and it takes rather small values, of 0.12–0.16 μF/cm^2, depending on $d_{\text{G-ch}}$ (95–155 nm) and W (55–67 nm).

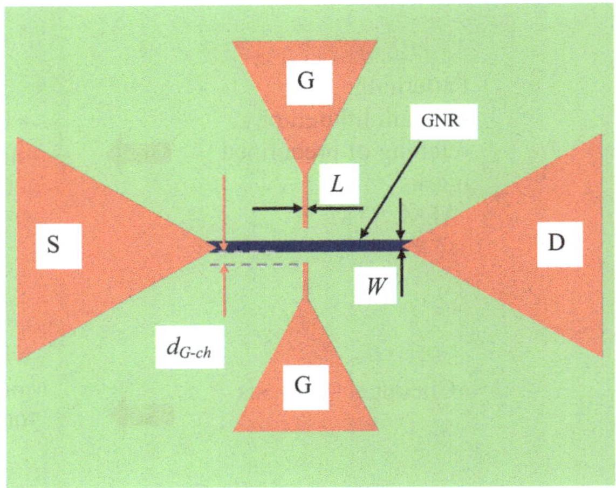

Fig. 2.30 Graphene FET with lateral gates

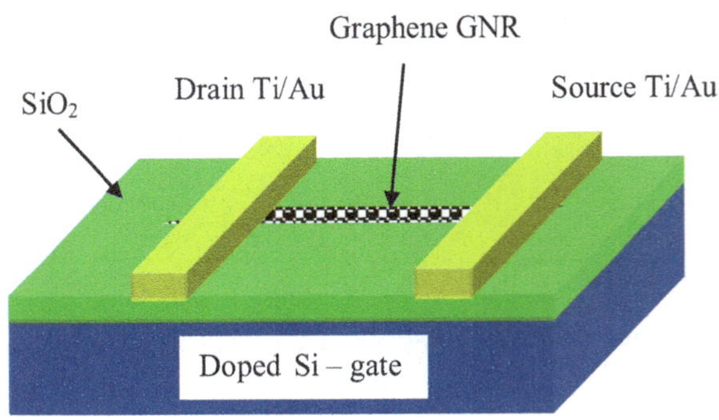

Fig. 2.31 Graphene GNR FET

A high transconductance is expected from side-gate transistors, similar or even better than the corresponding values of top-gate Si MOS FETs. Using a chemical method, sub-10 nm GNR widths were obtained and a first major result was a FET with a back gate, where 2 nm wide GNRs were transferred on a 10 nm thick SiO_2 grown on doped Si (Wang et al. 2008) (see Fig. 2.31).

The transistor has an *on*/*off* ratio of 10^6 at $V_D = 0.5\,\text{V}$, a width of 2 nm and a length of 230 nm, the mobility being of 200 $\text{cm}^2/\text{V s}$, with a mean free-path of carriers of 10 nm. The opening of the bandgap in graphene due to the lateral confinement of charge carriers produces a high *on*/*off* ratio, but the mobility of graphene decreases below that of a Si CMOS transistor. This is due to the scattering at the edges of the FET.

However, the above transistor cannot be replicated at the wafer scale. Wafer-scale GNR FETs were fabricated via e-beam resists, which cover the graphene, followed by plasma etching to define the GNR. Nanoribbons with widths in the range of 10–12 nm show bandgaps in the range of 0.1–0.15 eV, and therefore the *on*/*off* ratio is decreasing dramatically, attaining 20–50 at room temperature, with values of 10^6 being obtained at low temperatures (Hwang et al. 2012, 2015).

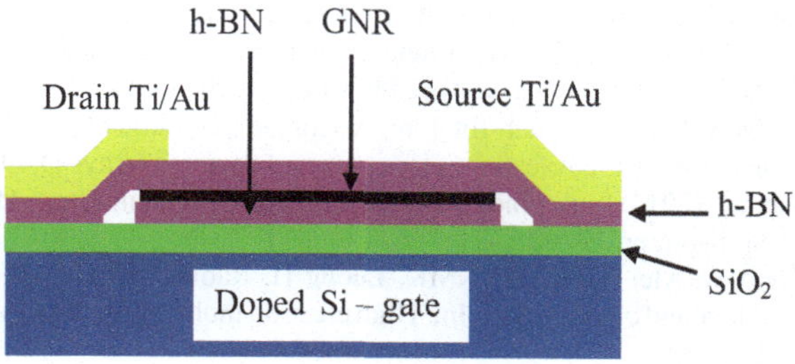

Fig. 2.32 GNR encapsulated between two h-BN monolayers FET

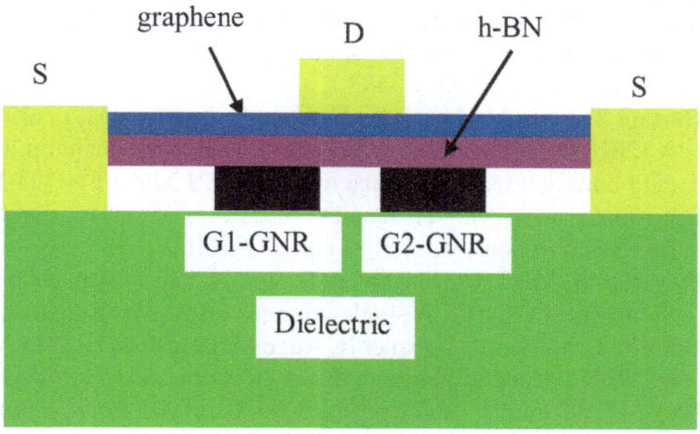

Fig. 2.33 Graphene GNR FET with h-BN dielectric in coplanar configuration for microwave applications

It is well known that h-BN improves dramatically the mobility of graphene, and especially of GNRs, when it is used as a substrate or if the GNR is sandwiched between two monolayers of h-BN (see Fig. 2.32). The h-BN induces a bandgap in graphene, and it is almost lattice matched with graphene. The mobility then reaches around 100,000 cm^2/V s, as discussed in the review (Somanchi et al. 2017). However, there are many difficulties to fabricate such a transistor at the wafer-scale, but the nanofabrication technologies are evolving very fast and one can expect in few years to see more atomic electronic devices at the wafer scale.

As an example, a microwave transistor based on GNRs was fabricated using h-BN as gate dielectric and having two metallic GNR back gates on the dielectric (Benz et al. 2013), denoted as G1 and G2 in Fig. 2.33. The channel of the transistor is graphene, and the drain (D) and the two source (S) electrodes form a coplanar line. The transistor has a cutoff frequency of 30 GHz.

References

Aono M, Hasegawa T (2010) The atomic switch. Proc IEEE 98:2228–2236

Banno N, Sakamoto T, Iguchi N, Kawaura H, Kaeriyama S, Mizuno M, Terebe K, Hasegawa T, Aono M (2006) Solid-electrolyte nanometer switch. IEICE Trans Electron 11:1492–1498

Benz C, Thurmer M, Wu F, Ben Aziza Z, Mohrmann J, Lohneysen H v, Watanabe K, Taniguchi T, Danneau R (2013) Graphene on boron nitride microwave transistors driven by graphene nanoribbon back-gates. Appl Phys Lett 102:033505
Celis A, Nair MN, Taleb-Ibrahimi A, Conrad EH, Berger C, de Heer WA, Tejeda A (2016) Graphene nanoribbons: fabrication, properties and devices. J Phys D 49:143001
Chua L (2014) If it's pinched it's a memristor. In Tetzlaff R (ed) Memristors and memristive systems. Springer, pp 17–90
Deal W, Mei XB, Leon KMK, Leong H, Radisic V, Sarkozy S, Lai R (2011) THz monolithic integrated circuits using InP High electron mobility transistors. IEEE Trans Terahertz Sci Technol 1:25–32
Dragoman M, Aldrigo M, Adam G (2017) Phased antenna arrays based on non-volatile resistive switches. IET Microwaves Antennas Propag 11:1169–1173
Dubois V, Raja SN, Gehring P, Caneva S, van der Zant HSJ, Niklaus F, Stemme G (2018) Massively parallel fabrication of crack-defined gold break junctions featuring sub-3 nm gaps for molecular devices. Nat Commun 9:3433
Durrani Z, Jonas M, Abualnaja F, Wang C, Kaestner M, Lenk S, Lenk C, Rangelow IW, Andreev A (2018) Room temperature single dopant atom quantum dot transistors in silicon formed by field-emission scanning probe microscopy. J Appl Phys 124:144502
Fang T, Konar A, Xing H, Jena D (2007) Carrier statistics and quantum capacitance of graphene sheets and ribbons. Appl Phys Lett 91:092109
Fuechsle M, Miwa JA, Mahapatra S, Ryu H, Lee S, Warschkow O, Hollenberg LCL, Klimeck G, Simmons MY (2017) A single-atom transistor. Nat Nanotechnol 7:242–246
Geng Z, Hähnlein B, Granzner R, Auge M, Lebedev Davydov VY, Kittler M, Pezoldt J, Schwiertz F (2017) Graphene nanoribbons for electronic devices. Ann Phys 529:1700033
Han MY, Özyilmaz B, Zhang Y, Kim P (2007) Energy Band-Gap Engineering of Graphene Nanoribbons, Phy Rev Lett 98:206805
Haemori M, Nagata T, Chkyow T (2009) Impact of Cu electrodes on switching behavior in a $Cu/HfO_2/Pt$ structure and resultant Cu ion diffusion. Appl Phys Express 2:061401
Hähnlein B, Händel B, Pezoldt J, Töpfer H, Granzner R, Schwierz F (2012) Side-gate graphene field-effect transistors with high transconductance. Appl Phys Lett 101:093504
Hamid E, Moraru D, Tarido JC, Miki S, Mizuno T, Tabe M (2010) Single-electron transfer between two donors in nanoscale thin silicon-on-insulator field-effect transistors. Appl Phys Lett 97:262101
He C, Shi Z, Zhang L, Yang W, Yang R, Shi D, Zhang G (2012) Multilevel resistive switching in planar graphene/SiO_2 nanogap structures. ACS Nano 6:4214–4221
Hwang WS, Tahy K, Li X, Xing HG, Seabaugh AC, Sung CY, Jena D (2012) Transport properties of graphene nanoribbon transistors on chemical-vapor-deposition grown wafer-scale graphene. Appl Phys Lett 100:203107
Hwang WS, Zhao P, Tahy K, Nyakiti LO, Wheeler VD, Myers-Ward RL, Eddy CR Jr, Gaskill DK, Robinson JA, Haensch W, Xing HG, Seabaugh A, Debdeep J (2015) Graphene nanoribbon field-effect transistors on wafer-scale epitaxial graphene on SiC substrates. Appl Phys Mater 3:011101
Jo SH, Chang T, Ebong I, Bhadviya BB, Mazumder P, Lu W (2010) Nanoscale memristor device as synapse in neuromorphic systems. Nano Lett 10:1297–1301
Koch M, Keizer JG, Pakkiam P, Keith D, House MG, Peretz E, Simmons MY (2019) Spin read-out in atomic qubits in an all-epitaxial three-dimensional transistor. Nat Nanotechnol 14:137–140
Martin CA, Smit RHM, van der Zant HSJ, van Ruitenbeek JM (2009) A nanoelectromechanical single-atom switch. Nano Lett 9:2940–2945
Moraru D, Udhiarto A, Anwar M, Nowak R, Jablonski R, Hamid E, Tarido JC, Mizuno T, Tabe M (2011) Atom devices based on single dopants in silicon nanostructures. Nano Res Lett 6:479
Pi S, Ghadiri-Sadrabadi M, Bardin JC, Xia Q (2015) Nanoscale memristive radiofrequency switches. Nat Nanotehnol 6:7519

Schindler C, Weides M, Kozicki MN, Waser R (2008) Low current switching in Cu-SiO_2 cells. Appl Phys Lett 92:122910

Somanchi S, Terrés B, Peiro J, Staggenborg M, Watanabe K, Taniguchi T, Beschoten B, Stampfe C (2017) From diffusive to ballistic transport in etched graphene constrictions and nanoribbons. Ann Phys 529:1700082

Strukov DB, Snider GS, Stewart DR, Williams RS (2008) The missing memristor found. Nature 453:80–83

Suga H, Horikawa M, Odaka S, Miyazaki H (2010) Influence of electrode size on resistance switching effect in nanogap junctions. Appl Phys Lett 97:073118

Suga H, Horikawa M, Kumaragurubaran S, Furuta S, Masuda Y, Shimizu T, Naitoh Y (2012) Resistance switch using metal nanogap electrodes in air. J Appl Phys 112:044309

Wang X, Ouyang Y, Li X, Wang H, Guo J, Dai H (2008) Room-temperature all-semiconducting sub-10-nm graphene nanoribbon field-effect transistors. Phys Rev Lett 100:206803

Xia Q (2014) Memristor device engineering and CMOS integration for reconfigurable logic applications. In: Tetzlaff R (ed) Memristors and memristive systems. Springer, pp 195–221

Xie F, Maul R, Augenstein A, Obermair Ch, Starikov EB, Schon G, Schimmel T, Wenzel W (2008) Independently switchable atomic quantum transistors by reversible contact reconstruction. Nano Lett 8:4493–4497

Xie F, Peukert A, Bender T, Obermair T, Wertz F, Schmieder P, Schimmel T (2018) Quasi-solid-state single-atom transistor. Adv Mat 30:1801225

Yang Y, Gu C, Li J (2019) Sub-5 nm metal nanogaps; physical properties, fabrication methods, and device applications. Small 15:1804177

Zwanenburg FA, Dzurak AS, Morello A, Simmons MY, Hollenberg LCL, Klimeck G, Rogge S, Coppersmith SN, Eriksson MA (2013) Silicon quantum electronics. Rev Mod Phys 85:961–1012

Chapter 3
Nanoelectronic Devices Enriching Moore's Law

Abstract This chapter is focused on nanoelectronic devices developed in the last years as the result of searching for alternative developments of the Moore's law. We deal in this chapter with ballistic devices, negative capacitance FETs, hyper FETs, tunneling devices, phase change devices, quantum dots and memories. Many of them have in common the fact that quantum mechanics is at the foundation of their functionalities, i.e., they are quantum devices.

3.1 FETs for Enhancing Moore's Law

3.1.1 Ballistic Transistors and Devices

In semiconductors, electrons, as charge carriers, are moving due to applied electrical fields and are scattered by impurities and phonons. As a result of various scattering effects, the carriers' mobility is reduced and heat dissipation occurs, which is the main adverse effects of the Moore's law. In CMOS technology, low mobility means a limited clock speed, while heat dissipation is limiting drastically the integration of more electron devices on the same chip area.

The Moore's law (Moore 1965), which is a visionary prediction, a road-map put forward more than 50 years ago, forecasts that the number of transistors on the same chip area will double each 18 months, and thus the dimensions of transistors will downscale with 50% in the same duration of time. However, due to the above-mentioned physical constrains, i.e., reduced mobilities and heat dissipation, the Moore's prediction is slowing down/is becoming too optimistic.

One alternative to surpass the main obstacles in front of the Moore's law, i.e., low mobility and heat dissipation, is the ballistic transport of charge carriers. The ballistic transport is defined as the collisionless transport regime, which occurs only when the dimensions of the sample are smaller than the mean free path L_{fp}. In the ballistic transport regime, the carriers behave as matter waves, which is a direct manifestation of quantum mechanics (Datta 1997), so that they have the highest possible velocity, which is about $c/300$ in the case of graphene monolayers. Due to the quantum nature of ballistic transport, the Ohm's law is replaced by the Landauer's law:

M. Dragoman and D. Dragoman, *Atomic-Scale Electronics Beyond CMOS*,
https://doi.org/10.1007/978-3-030-60563-6_3

$$I = \frac{2e}{h}\int M(E)T(E)[f_l(E) - f_r(E)]\mathrm{d}E \tag{3.1}$$

where I is the current, E is the energy of electrons, $f_l(E)$ and $f_r(E)$ are the Fermi-Dirac quasi-distribution functions in the two (left and right, respectively) metallic contacts between which a voltage V is applied. Further, M represents the number of transverse modes and T is the quantum probability of an electron injected in one contact to be transmitted to the other. The dependence $T(E)$, which contains information on the potential energy and hence the voltage applied between the contacts, is directly connected to the quantum nature of the ballistic transport, being the bridge between two macroscopic quantities: input voltage V and output current. Thus, the ballistic transport regime is governed by quantum mechanics, and deals with coherent charge carriers/wavefunctions, while the diffusive transport is incoherent. Therefore, the heat dissipation in the ballistic transport is minimized as much as possible.

The mean free path L_{fp} is a measure of the distance between collisions of carriers with impurities or phonons and dictates the limits of the ballistic transport. In Table 3.1 we have displayed the parameter L_{fp} at room temperature for several nanomaterials. This table shows that graphene displays the best L_{fp} at room temperature, although the first ballistic FETs were designed and tested using semiconductor heterostructures.

Plasma of charge carriers in 2D gas is used to model ballistic transport in FETs working at very high frequencies, reaching THz (Lusakowski et al. 2005; Knap et al. 2008). Among ballistic devices we mention the ballistic rectifier (BR), which is an asymmetric four-terminal cross-junction based on ballistic carrier transport, able to rectify alternative currents. Such rectifiers were implemented as a junctionless diode, used as an InGaAs/InAlAs/InP deflection transistor (Diduck et al. 2009), or in graphene (Auton et al. 2017). Thus, the graphene monolayer has become the most desirable material for ballistic FETs.

Table 3.1 L_{fp} at room temperature

Material	L_{fp} at room temperature
InSb/AlInSb quantum well	580 nm
Graphene	500 nm, in good quality graphene monolayer
	>1 μm for graphene encapsulated in BN monolayers
	>10 μm in graphene nanoribbons
InAlN/GaN on SiC	75 nm
InAs nanowire	150 nm
Carbon nanotubes	>1 μm
Si	2 nm

The first ballistic graphene FET at room temperature was designed to remedy some unsolved issues related to graphene FETs. First, a nonlinear drain current-drain voltage ($I_D - V_G$) characteristic, useful in many applications such as detection, mixing and even amplification, was intended, since the non-ballistic graphene monolayers FETs have a linear $I_D - V_G$ dependence, acting as voltage-controlled resistances. Second, the amplification, which is a parameter specific for any transistor, was desired, taking into account that non-ballistic graphene monolayer FETs show no amplification due to the lack of the saturation region, i.e., the lack of a bandgap. As a consequence, in graphene non-ballistic transistors the ratio between transconductance g_m and drain conductance is less than 1.

On the other hand, (3.1) represents a generally nonlinear dependence between current and voltage in the ballistic regime. Therefore, if a ballistic regime occurs in the graphene channel, the curve $I_D - V_G$ could be nonlinear and dependent on V_G, so that it can even saturate at high current densities.

However, from the point of view of applications in nanoelectonics, there is a detrimental effect in graphene ballistic transport, not encountered in any other 2D materials, due to the Dirac-like transport equation of charge carriers. Namely, the unique Dirac-like transport in graphene generates the so-called Klein paradox. The Klein paradox manifests itself in the lack of electrostatic modulation of the channel at normal incidence of ballistic carriers, and originates in the fact that the transmission probability is always 1 in this situation, irrespective of the gate potential, i.e., $T = 1$. Therefore, in order for the transmission to be modulated between 0 and 1 (and hence, for the current to be controlled) by the gate voltage, the ballistic FET in graphene must have oblique gates, for instance at 45° with respect to the drain (D)-source (S) axis. In this way, the gate control of ballistic charge carriers mimics the effect of a bandgap, and allows the ballistic transistor to be in *on* and *off* states, similar to as semiconductor transistor (see Fig. 3.1(a)).

In a graphene monolayer FET with oblique gate, in a certain bias range, $T(E)$ is abruptly decreasing to zero as the voltage is increasing, such that the current experiences a negative differential resistance (NDR) region in the transition between the *on* and *off* states. This NDR region can be seen as an imprint of the ballistic transport regime in graphene transistors (see Fig. 3.1(b)) (Dragoman and Dragoman 2007a). Unlike in semiconductor devices, where NDR is due to tunneling through a discrete energy level (and hence to increased transmission of charges from one end of the device to another), in graphene FETs it is caused by a diminishing of transmission probability of charge carriers.

The ballistic graphene FET with obliqueed gates was fabricated on a 4 inch wafer of graphene monolayer, which was previously transferred on a Si/SiO_2 wafer. The gate insulation was HSQ, with a thickness of 50 nm, and the distance between D and S electrodes was 400 nm. The gate length was 40 nm (Dragoman et al. 2014a). This ballistic FET has shown NDR regions in the $I_D - V_G$ dependence at room temperature, with a peak-to-valley ratio (PVR) of 9, while the top gate has changed the NDR, modifying the PVR in the range 7–9. In Fig. 3.2 we have represented the image of the ballistic graphene FET.

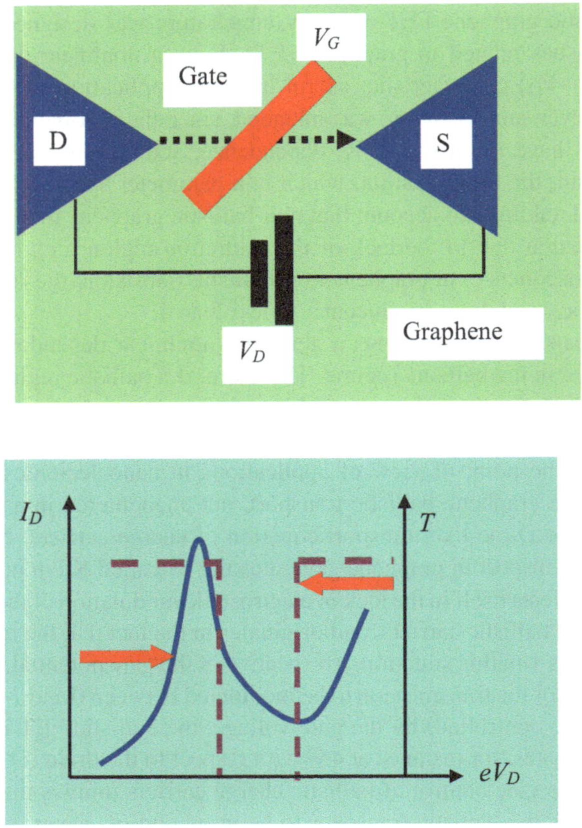

(a)

(b)

Fig. 3.1 (a) Ballistic FET based on graphene monolayers with tilted gate, and (b) transmission and current signatures of ballistic graphene FET

A further development of the above device consisted of the design and fabrication of a multi-gate ballistic FET (Dragoman et al. 2016a) working at room temperature. We have used two-tilted gates at 45°, with a gate length of 30 nm and a drain-source distance of 190 nm. The SEM image of this ballistic transistor is given in Fig. 3.3. The rectangular dark area in this figure represents the dielectric HSQ, with a thickness of 40 nm, over which the two oblique gates, which can be individually polarized, were patterned. The HSQ stands over the graphene channel.

The 4-in. graphene monolayer/SiO_2/Si wafer was transformed into 70 double-gate FETs after twelve different fabrication steps based on electron beam lithography (EBL), reactive ion etching (RIE), metal deposition, and lift-off (see Fig. 3.4). More precisely, the graphene channels are patterned by EBL and then etched by RIE in oxygen plasma, being precisely placed in a set of previously fabricated register marks (see Fig. 3.4(a), (b)). The S and D contacts are patterned with a very thin (50 nm) PMMA (poly(methyl methacrylate)) layer and then the metallic contacts are

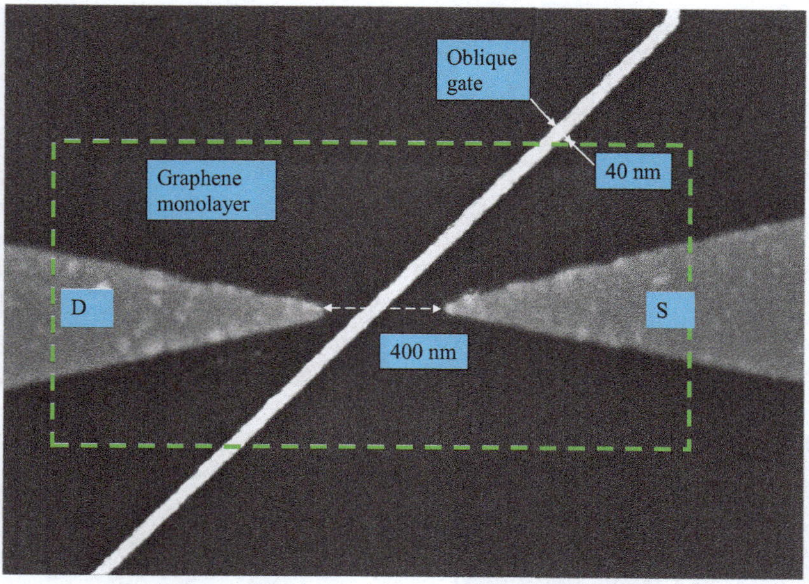

Fig. 3.2 SEM image of the ballistic graphene FET (Dragoman et al. 2014a)

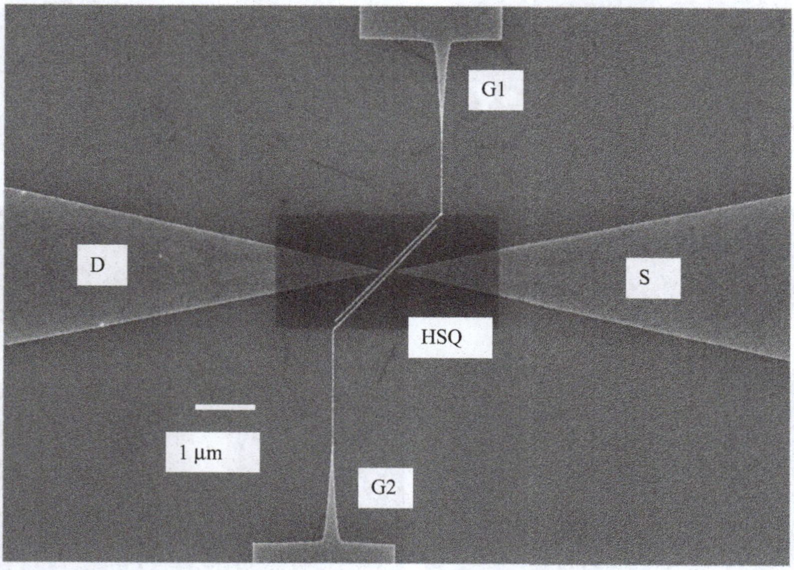

Fig. 3.3 SEM image of the double-gate ballistic graphene FET (Dragoman et al. 2016a)

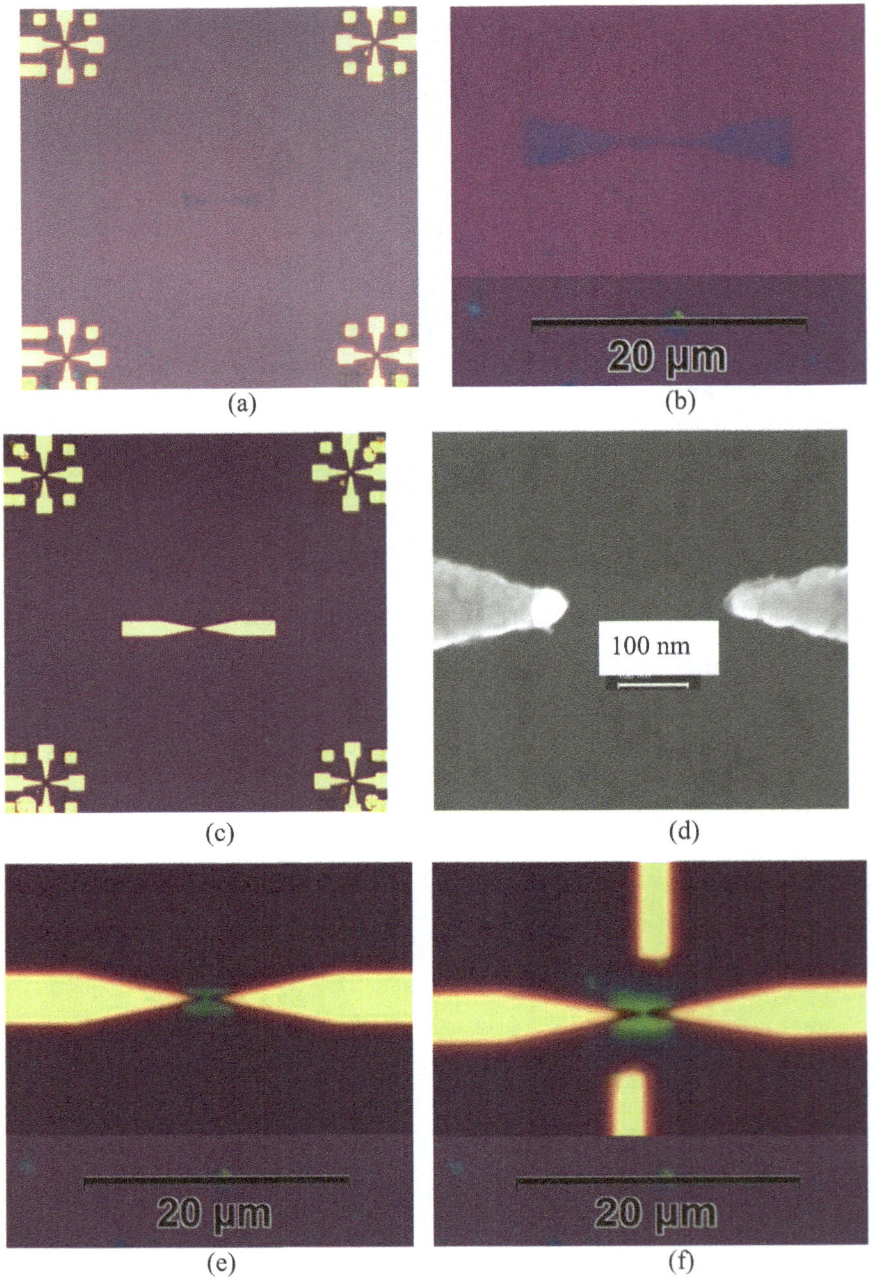

Fig. 3.4 The optical images of various steps of fabrication oh the ballistic double gate graphene FET (Dragoman et al. 2016a)

fabricated via electron beam evaporation using Ti/Au (7 nm/30 nm) (see Fig. 3.4(c), 3.4(d)). The FET fabrication has continued by dielectric deposition over the graphene channel. In this case, the dielectric was a very thin negative electron resist (HSQ-XR1541-2). The dielectric with a thickness of 40 nm was deposited by spinning and then shaped by EBL (Fig. 3.4(e)). Finally, the electrodes of the two tilted gates are fabricated. The electrodes are patterned in PMMA and then fabricated by lift-off from Ti (7 nm) and Au (40 nm) (Fig. 3.4(f)). The gate lengths are 30 nm and the separation between them is of 40 nm.

The $I_D - V_G$ dependence of the graphene FETs is strongly nonlinear at various gate voltages. This dependence shows a saturation region and a NDR behavior at some gate voltages. The maximum transconductance, g_m, is 1.42 mS/μm at $V_D =$ 3 V and $V_{TG} = -0.5$ V, a very high value for graphene FETs. The NDR can be observed in the drain conductance, g_D, behavior (see Fig. 3.5(a)). More precisely, when g_D is decreasing very much, the ratio between the transconductance g_m (see Fig. 3.5(b)) and g_D, g_m/g_D, is higher than 50 for a drain voltage within the range of 1.5–2.5 V, so we have voltage amplification. This is a signature of ballistic transport, since graphene FETs are not amplifying in the drift-diffusion regime because g_m/g_D is generally less than 1.

Ballistic devices have important applications in quantum computing. Quantum computing is a hot topic of research, and the first quantum computers are currently built by very large companies using superconducting circuits. These computers are very bulky and work at few mK. The main problem of quantum computing is the loss of coherence of the quantum wavefunction, which destroys the quantum computation results. Ballistic devices could offer a solution to this problem since the wavefunction of ballistic carriers are coherent up to device dimensions attaining the mean free path, and the imperfections of the edge of ballistic devices are very small. We have designed and fabricated one-qubit Hadamard or NOT quantum gates (QG) and a one-qubit modified Deutsch-Jozsa (DJ) quantum algorithm circuit on 4 in. wafer of graphene monolayer transferred on a Si/SiO_2 wafer and we have demonstrated the quantum coherence of the device via current measurements (Dragoman et al. 2018a).

The Hadamard quantum gate, denoted by QG in Fig. 3.6, consists of a Y-shaped junction designed to support a single mode, created by the superposition of $|0\rangle$ and $|1\rangle$ quantum logic states, which are the wavefunctions of the upper and lower branches of the Y-junction, respectively. The output of the Y-junction is $(|0\rangle + |1\rangle)/\sqrt{2}$. After this junction, there is an interference region of length L design especially to support only two transverse modes. The two output currents measured at the end of the interference region are depending on the phase difference between the propagating modes in this region, given by:

$$\Delta\phi(E, U_G, L) = L\left[\sqrt{(E-U_G)^2/(\hbar v_F)^2 - (2\pi/W)^2} - \sqrt{(E-U_G)^2/(\hbar v_F)^2 - (\pi/W)^2}\right]. \tag{3.2}$$

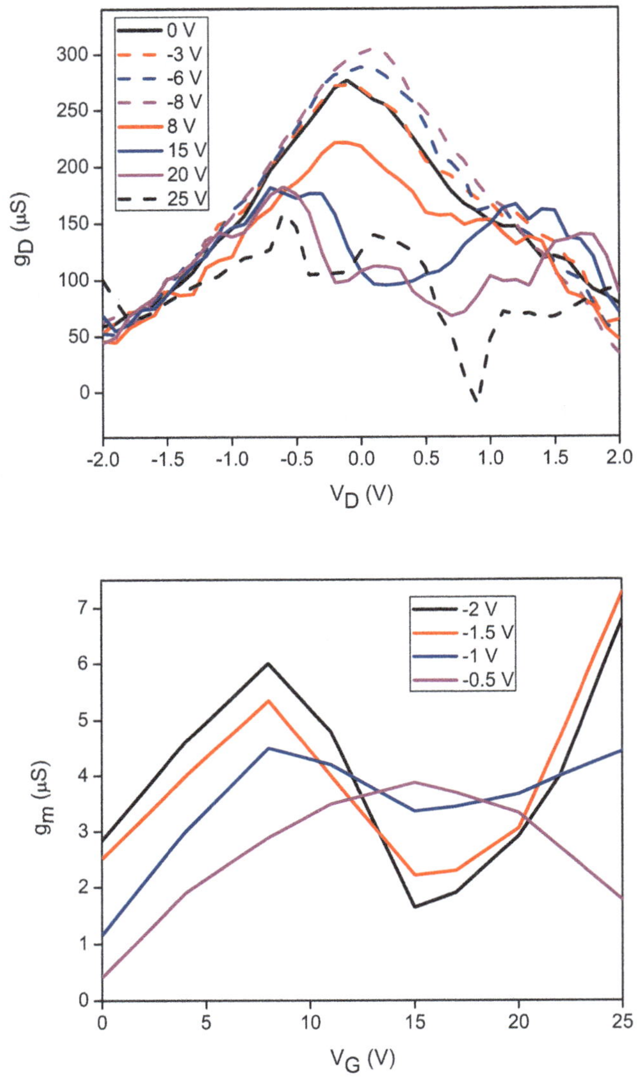

(a)

(b)

Fig. 3.5 (a) The drain conductance dependence on drain voltage at several gate voltages applied on G_1 indicated in the legend, and (b) the transconductance dependence on the gate voltage on G_1 at various drain voltages given in the legend. The gate voltage on G_2 is -3 V in all cases

If the transport is not ballistic, the ratio between the two output currents is independent of the back-gate voltage and no coherence is present. On the contrary, we have observed that the ratio between the output currents is dependent on the back-gate voltage by performing current measurements, as indicated in Fig. 3.7. So, the quantum coherence is evidenced.

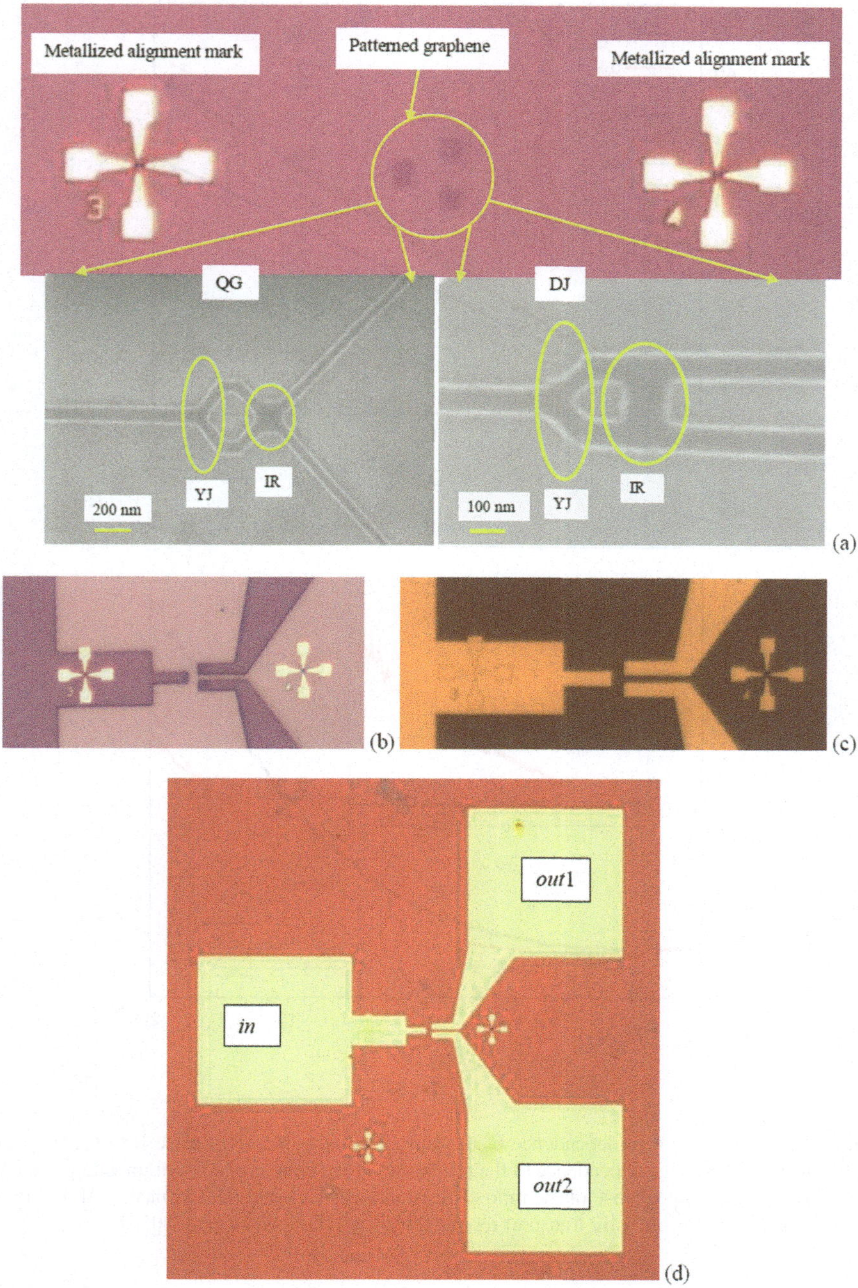

Fig. 3.6 Quantum graphene devices fabrication: (a) graphene patterning of QG or DJ devices, (b) patterning and (c) metallization of the contact pads, and (d) the final three-terminal device (Dragoman et al. 2018a)

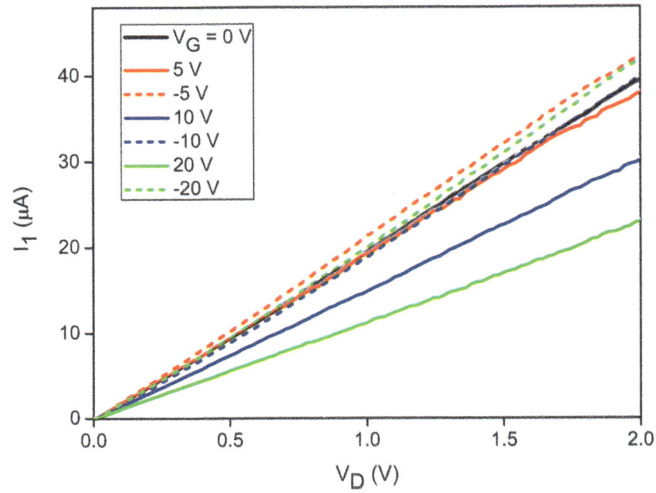

(a)

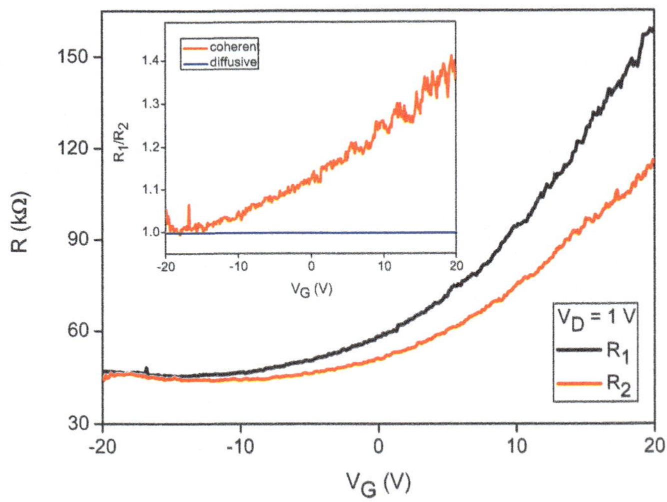

(b)

Fig. 3.7 (a) Drain voltage dependence of the current at *out*1 in a QG device for different gate voltages (b) Gate voltage dependence of the resistances at *out*1 and *out*2 for a drain voltage of 1 V. Inset: gate voltage dependence of the ratio of resistances at *out*1 and *out*2 as measured (red line) and as expected in the diffusive transport regime (blue line) (Dragoman et al. 2018a)

The ballistic transport is used also in advanced Josephson junctions (JJ) for quantum computing. A JJ is a superconductor-insulator-superconductor (SIS) structure, where S are metals. In superconductors, the carriers are coherent, i.e., a single wavefunction describes their movement, and a unique phase Φ is the phase difference between the two superconductors. In the insulator, the carriers are tunneling and preserve the phase, so the current of JJs depend on the phase, as in the case of quantum gates based on graphene. Weak links are another configuration of JJs where the metals are replaced by nanoconstrictions.

The JJ is ballistic if its length L is smaller than L_{fp} (Lee et al. 2015). JJs devices involving graphene are designed in two-main configurations: (i) with in-plane electrodes, and (ii) with vertical electrodes (see Fig. 3.8). Both configurations have some disadvantages (Li et al. 2018). In the first graphene JJ configuration, the roughness of metallic electrodes is small. Large roughness destroys the ballistic transport. The second configuration, where graphene is located between two metals and covered with h-BN to reduce the scattering between graphene and electrodes, is difficult to be fabricated and the reproducibility at wafer level is problematic.

Using ballistic JJs, more complex devices, such as tunable cavities, can be designed and fabricated. An example is the gate tunable microwave cavity (Schmidt et al. 2018), illustrated in Fig. 3.9. The ballistic JJ is inserted into a coplanar cavity, which is tuned from about 7.1–8.2 GHz with a top-gate voltage, the gate voltage being varied in this case between −30 V and 30 V. The D (drain), S (source) and G (gate) electrodes form the coplanar waveguide; HSQ is the gate dielectric. The length of the ballistic JJ is 500 nm and its width is of 5 μm.

The microwave cavity involves a coplanar line (CPW) configuration formed from three metallic electrodes, two of them grounded (G) and encompassing the signal electrode S, which is the gate, decoupled by a capacitance from the microwave signal. The tunable element is the ballistic JJ inductance, given by:

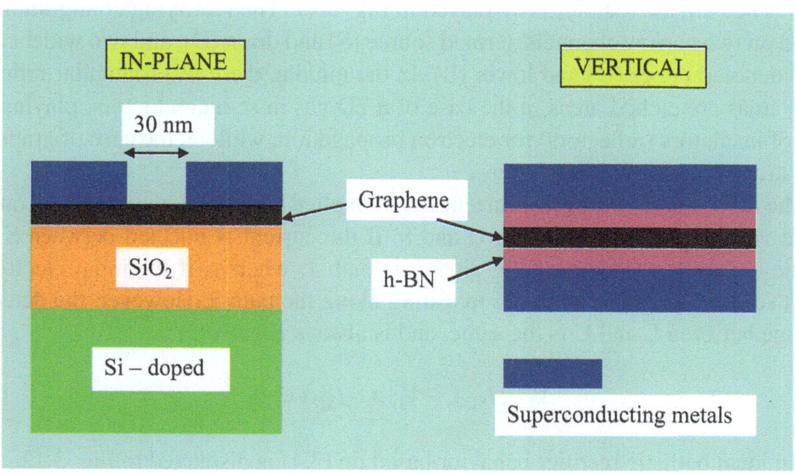

Fig. 3.8 Ballistic Josephson graphene junctions

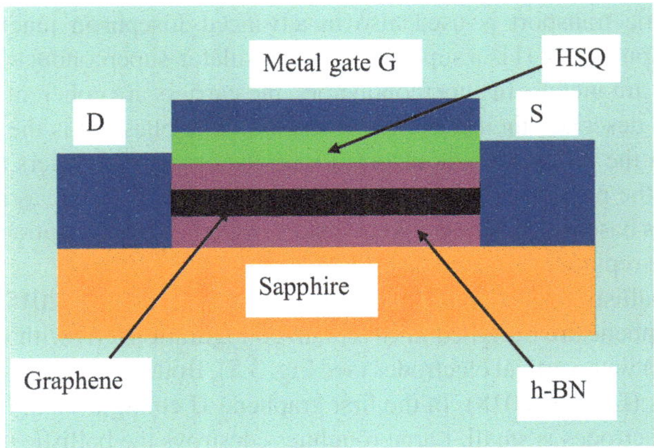

Fig. 3.9 DC tunable microwave cavity based on ballistic JJ

$$L_{JJ} = I_C \Phi / 2\pi \tag{3.3}$$

By changing the critical current I_C via the gate voltage, the JJ inductance is changed.

Ballistic transistors are outperforming FETs based on Si or GaAs. An example is a FET formed from parallel a carbon nanotubes (CNTs) array where the density of CNTs is 47 CNTs/μm. The channel length is 150 nm and its width is 4 μm (see Fig. 3.10) (Brady et al. 2016). The ballistic CNT FET has in the saturation, *on* state, a density of current of 900 μA/μm, which exceeds the values of Si or GaAs FETs. The transconductance is 1.7 mS/μm at room temperature.

One of the most studied ballistic devices is the ballistic rectifier (Song 2004). Initially, it was implemented using a 2D electronic gas in semiconductor heterostructures. The ballistic rectifier is displayed in Fig. 3.11. The rectifying configuration is based on two narrow channels, termed source (S) and drain (D), and two wider channels, denoted upper (U) and lower (L). In the middle, there is a triangular reflector Blue areas are etched areas in the case of a 2D gas in semiconductors, playing the role of insulators (reflectors) for electron propagation, while in the case of graphene blue areas are metallic electrodes.

The input signal, which is a current I_{SD}, is applied between S and D and the output is the voltage collected between U and S. If the current is injected between S and D, the electrons are propagating along the path 1, whereas if the input electrodes are reversed, the electrons are propagating along the path 2. However, the detected voltage between L and U is the same, and is always negative, i.e.,

$$V_{LU}(I_{SD}) - V_{LU}(-I_{SD}) = 0. \tag{3.4}$$

An ideal ballistic rectifier behavior based on (3.4) is displayed in Fig. 3.12. Denoting $R_{SD,LU} = V_{LU}/I_{DS}$, the rectifier satisfies the relation

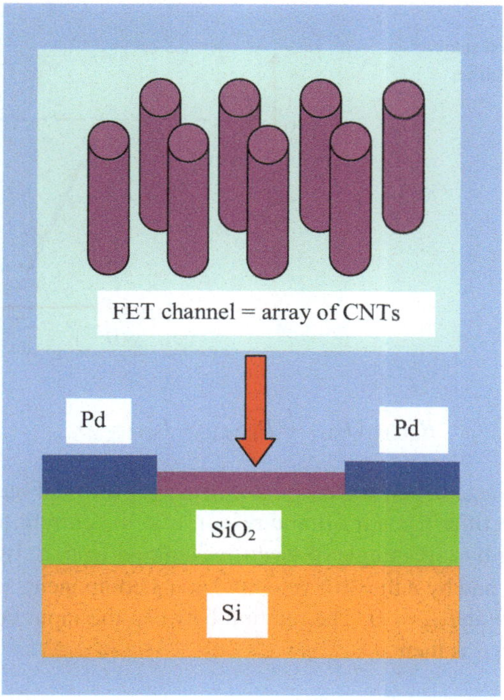

Fig. 3.10 Ballistic CNT array FET

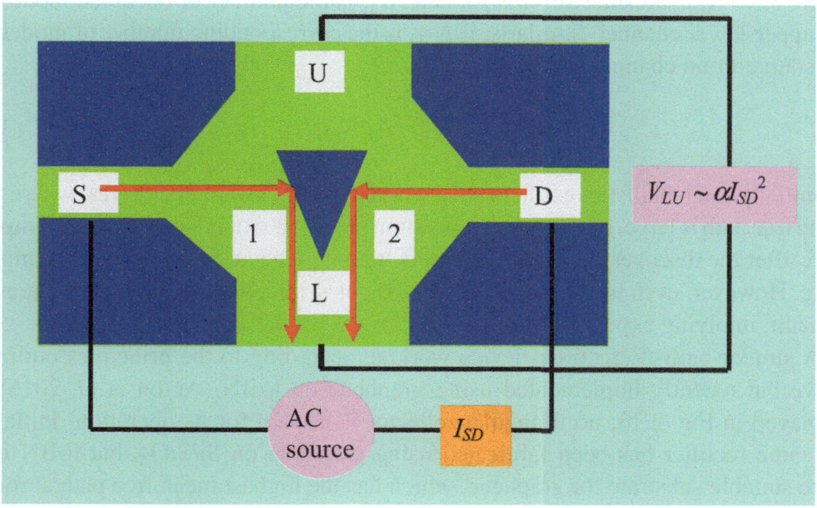

Fig. 3.11 Ballistic rectifier

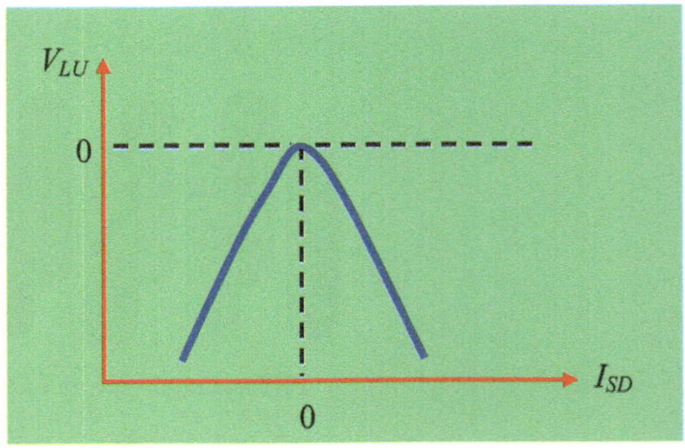

Fig. 3.12 Typical current-voltage characteristic of an ideal ballistic rectifier

$$R_{SD,LU}(I_{SD}) + R_{SD,LU}(-I_{SD}) = 0. \quad (3.5)$$

The current-voltage dependence of the rectifier is given in Fig. 3.12.

The ballistic rectifier has a quadratic response to the current, i.e., is a quadratic detector. Denoting by θ the angle of electrons emitted from S, by θ_0 the minimum angle from S to D, and by v the drift velocity having components v_x and v_y, we have $\theta = \arctan(v_y/v_x)$ at $I_{SD} = 0$. The relation between the input current I_{SD} and the rectified voltage V_{LU} is then

$$V_{LU} = -(3h/4e^2 N_{LU})(\sin\theta_e - \sin\theta_0)|I_{SD}|, \quad (3.6)$$

where $\theta_e = \theta_0 + \arcsin(v \sin\theta_0/v_F)$, and N_{LU} is the number of occupied modes in the upper-lower channel. Similarly, if N_{SD} is the corresponding number of modes in the source-drain channel, we have

$$V_{LU} = -(3\pi\hbar^2/4e^3 E_F N_{LU} N_{SD}) I_{SD}^2. \quad (3.7)$$

Initially, such ballistic rectifiers have worked only at low temperatures, where the mean free path is large, but later their operation range has extended up to room temperature after the discovery of semiconductors with high mean free path at this temperature. However, even today the growth of 2D gas semiconductor heterostructures is difficult, implying expensive equipments.

A similar ballistic rectifier device working according to the principles outlined above but recently implemented using graphene on h-BN (Auton et al. 2016) is displayed in Fig. 3.13; no triangular reflector is needed for its operation. Initially, the same rectifier has been fabricated using graphene on Si/SiO_2, but h-BN is a more suitable substrate for graphene, which has the highest mean free path at room temperature. Such a ballistic graphene rectifier can be used, together with an antenna, to detect signals with THz frequencies. The mobility of this detector is determined to be 200,000 cm^2/V s and its responsivity is of 23,000 V/W, with a noise-equivalent power of 0.64 pW $Hz^{-1/2}$, but these parameters are estimated at low frequencies. At THz frequencies, the responsivity reduces to about 760 V/W (Auton et al. 2017).

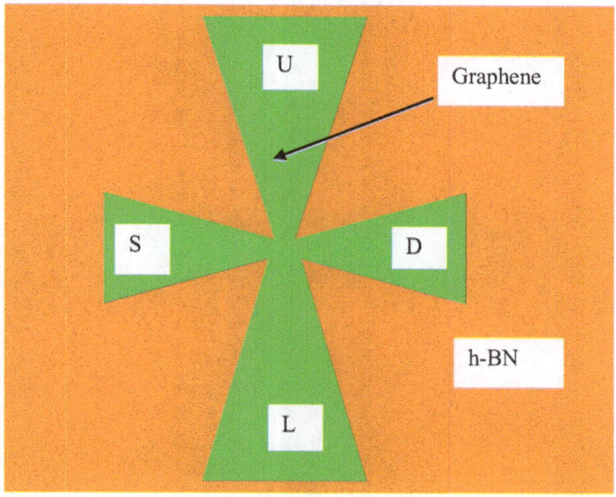

Fig. 3.13 Ballistic rectifier without a distinct reflector

As already mentioned before, the ballistic transport is enhancing the performances of FETs at high-frequencies, up to the THz region, and provides voltage gain, results almost impossible to obtain with graphene FETs working in the drift-diffusion regime. However, graphene suffers still from the lack of a bandgap. The absence of the bandgap means that graphene FETs cannot be closed or blocked, and thus the basic property of a FET, i.e., to be switched on and off, is lost. Note that the FET switch is the key component of any computer, smart phone and many other electronic products. Due to advancements in semiconductor technology, billions of FET switches are integrated in a single chip, which is programmable to perform various tasks. On the other hand, graphene FETs can be integrated on a single chip, but due to the defects the yield is low. An example is a high-frequency circuit, with an area of 0.6 mm^2, formed from a two-stage amplifier and a mixer stage. The circuit incorporates three graphene transistors, with gate lengths of 900 nm (Han et al. 2014). As a high-frequency circuit, this circuit has not very good performances, and this is largely due to the lack of bandgap, meaning also that no saturation region is attained.

There are many methods to create a bandgap in graphene, and one of them is to fabricate ultranarrow graphene nanoribbons, discussed in the previous chapter; the mobility of electrons in such nanoribbons and devices based on them is, however, very low. On the other hand, when graphene is transferred on certain substrates, an induced bandgap is created. In the case of graphene/h-BN, the theory predicts an induced bandgap of 0.5 eV (Giovannetti et al. 2007), but the FETs based on graphene/hBN show no blocking region, only high mobilities. This means that either h-BN does not induce any bandgap in graphene, or that there are misalignments in crystal orientations between h-BN and graphene. SiC induces a bandgap of $E_g =$ 0.26 eV in the epitaxially-grown graphene (Zhou et al. 2007). This small bandgap is enough to demonstrate an on-off FET on the Si-face of SiC, with clear saturation and *on-off* regions, a high transconductance, and a very high mobility of 6000 cm^2/V s (Moon et al. 2010).

Fig. 3.14 SEM image of the nanopatterned graphene channel (Dragoman et al. 2018b)

We have found recently that nanopatterned graphene channels are able to solve the so-called graphene electronics conundrum: the simultaneous existence of bandgap and high mobility of carriers (Dragoman et al. 2018b). Nanopatterning means the carving in the FET channel of a 2D array of holes with e-beam techniques, as shown in Fig. 3.14. The diameter of the holes are 20 nm, their array has a period of 100 nm, and are perforated in a graphene nanoribbon with a width of $W = 2\ \mu\text{m}$. This periodic structure induces a bandgap in graphene of about 0.2 eV, very close to the SiC-induced bandgap. Such holes are termed often also as antidots.

Nanopatterning of holes generally degrades the physical properties of the graphene monolayer via the introduction of defects and edge disorder. However, very high mobility and ballistic transport at room temperature with a carrier mean free path of 400 nm can still be preserved if the perforated graphene layer is sandwiched between h-BN layers (Sandner et al. 2015) or, as in our case, by using HSQ resist as top-gate dielectric.

We have fabricated nanopatterned graphene FETs with channel lengths of $L = 1$, 2, 4 and 8 μm and $W = 2\ \mu\text{m}$. Two SEM images of such a transistor are displayed in Fig. 3.15.

A typical current-drain voltage dependence of a nanopatterned graphene FET with $L = 2\ \mu\text{m}$, for different top-gate voltages, is displayed in Fig. 3.16. From this figure it follows that the transistor is blocked at $V_G = -0.2$ V.

The mobility can be extracted from experimental data based on the formula

$$\mu = g_m L / W C_g |V_D| \tag{3.8}$$

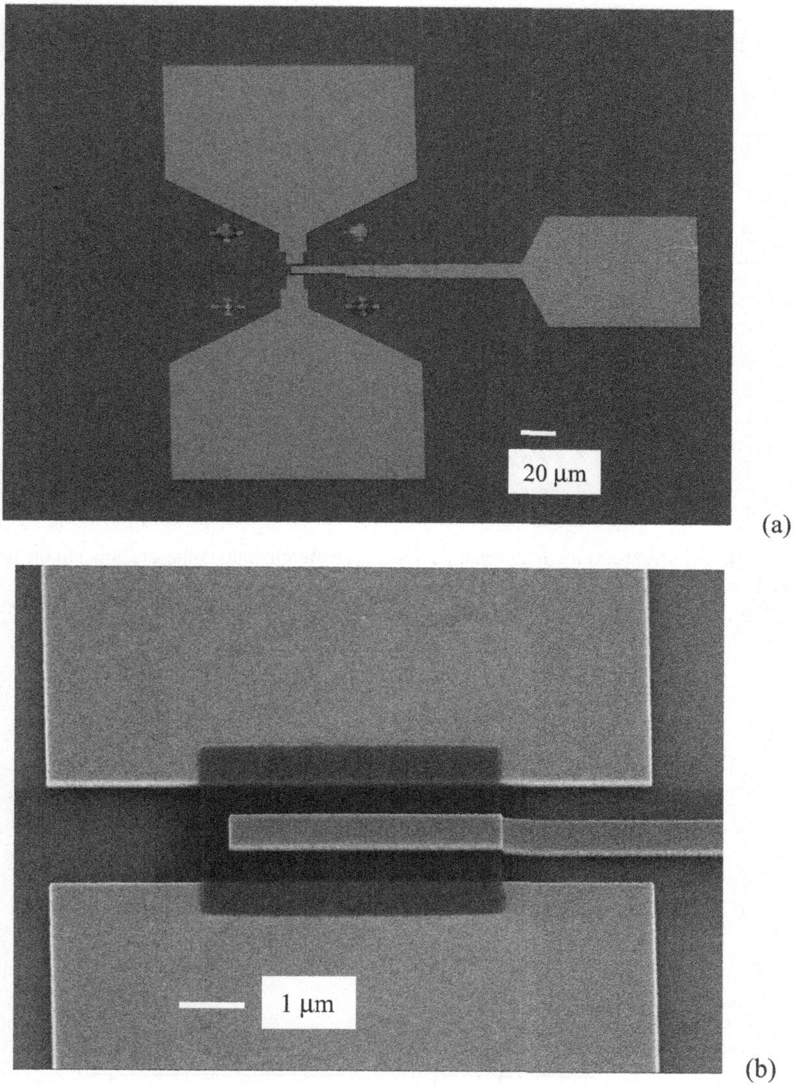

Fig. 3.15 (a) SEM image of the nanopatterned GFET, and (b) a corresponding detail of the source-drain region

where $g_m = \partial I_D / \partial V_G$ is the transconductance, L and W are the length and width, respectively, of the graphene nanopatterned channel, and $C_g = \varepsilon_0 \varepsilon_r / t$ is the gate capacitance per unit area, with t the thickness of the gate dielectric, ε_0 the vacuum permittivity and ε_r the relative dielectric permittivity.

The mobility dependence on the channel length in graphene FETs with nanopatterned channels is depicted in Fig. 3.17. As can be observed, the mobility in these 2 μm-wide transistors is ranging from 10,400 $cm^2/V\,s$ for a channel length of 1 μm to

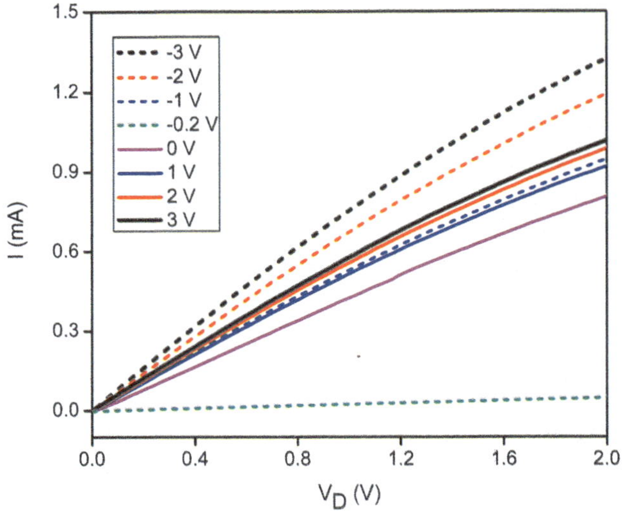

Fig. 3.16 Drain current-drain voltage characteristics at different gate voltages shown in the legend for devices with $L = 2\ \mu m$ (Dragoman et al. 2018b)

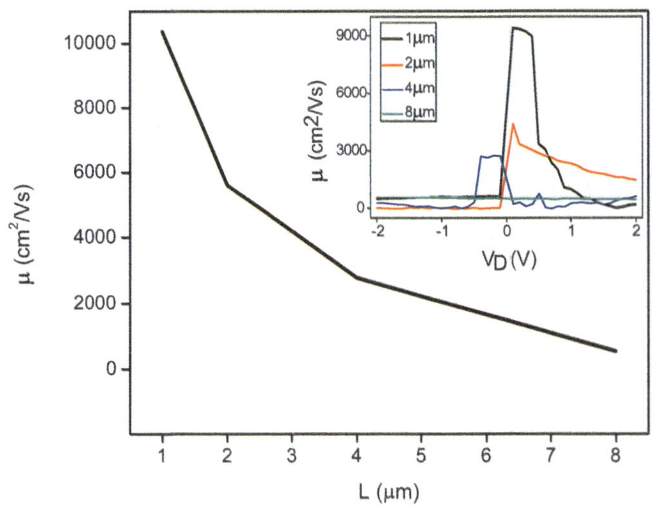

Fig. 3.17 Dependence of the mobility on the nanopatterned channel length. Inset: dependence of mobility on drain voltage at different channel lengths and for $V_G = 0.5$ V (Dragoman et al. 2018b)

about 550 cm^2/V s for a channel length of 8 μm. Irrespective of the mobility value, in all transistors the *on*/*off* ratio is at least 10^3. So, as a function of the length of graphene FETs, the measured mobilities are comparable to those of semiconductors

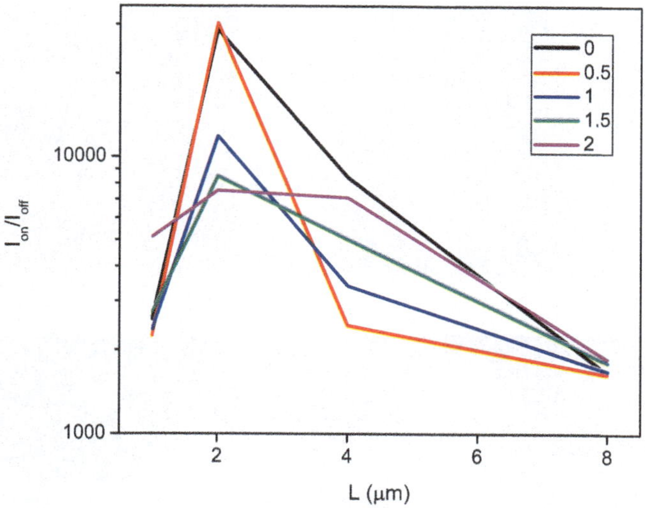

Fig. 3.18 The *on/off* ratio calculated between drain currents at $V_D = 1$ mV and 2 V versus channel length at gate voltages in the inset (Dragoman et al. 2018b)

such as InP, GaAs, GaN and Si. Thus, this powerful concept, of FETs with nanopatterned graphene channel, reduces the problems facing graphene integrated circuits to a single one: the growth of graphene monolayers at large scale with reduced defects.

Since nanopatterning is producing a bandgap, it is expected to obtain a large *on/off* ratio, which does not exist in graphene FETs which have a continuous, non-patterned channel. If $E_g = 0.16\,\text{eV}$, for instance, we expect an *on/off* ratio of $I_{on}/I_{off} \cong \exp(E_g/k_BT) \cong 500$, while for a larger bandgap, of $E_g = 0.2\,\text{eV}$, we get $\exp(E_g/k_BT) \cong 2400$, value which is retrieved from our experiments (see Fig. 3.18).

The fabrication technology of nanopatterned graphene FETs involves several steps, as indicate in Fig. 3.19 (Dragoman et al. 2017):

(i) patterning of the alignment marks, followed by e-beam evaporation of Cr (10 nm) and Au (100 nm) for metallization of alignment marks and lift-off,
(ii) patterning of the graphene layer (Fig. 3.19(a)),
(iii) reactive ion etching of graphene channel (Fig. 3.19(b)),
(iv) patterning of source and drain contacts (Fig. 3.19(c)),
(v) e-beam evaporation of Ti (10 nm) and Au (100 nm) followed by lift-off (Fig. 3.19(d)),
(vi) patterning the gate insulating layer (Fig. 3.19(e)),
(vii) patterning the metallic gate contact (Fig. 3.19(f)),
(i) (vi) e-beam evaporation of Ti (10 nm)/Au (100 nm) followed by lift-off (Fig. 3.19(g)),
(viii) dicing of the wafer e wire bonding (Fig. 3.19(h), (i)).

Fig. 3.19 Optical images of the structures obtained after several technological processes of a nanopatterned transistor (Dragoman et al. 2017)

3.1.2 2D-Materials Based Transistors

We have discussed extensively about transistors based on 2D materials, such as TMDs, phosphorene and other materials in (Dragoman and Dragoman 2017). Therefore, here we will discuss only the possible developments and applications of 2D-materials based transistors. There are four critical parameters for atomically thin transistors:

- bandgap and workfunction (Table 3.2)
- mobility and effective mass (Table 3.3) (from Dragoman et al. 2019)

Table 3.2 Bandgaps and workfunctions of some atomically thin materials (from Dragoman et al. 2019)

2D material monolayer	E_g (eV)	Type of gap	Workfunction (eV)
HfS_2	1.23	Indirect	−5.7
$HfSe_2$	0.45	Indirect	−5.17
MoS_2	1.6	Direct	−5.07
$MoSe_2$	1.31	Direct	−4.6
WS_2	1.54	Direct	−4.70
WSe_2	1.32	Direct	−4.2

Table 3.3 Mobilities and effective masses of some atomically thin materials (from Dragoman et al. 2019)

2D material	Mobility (cm^2/V s)	$m_{eff}\ \Gamma - K \times m_e$	$m_{eff}\ K - M \times m_e$
HfS_2	1833	3.30	0.24
$HfSe_2$	3579	3.10	0.18
MoS_2	340	0.45	0.45
$MoSe_2$	240	0.52	0.52
WSe_2	0.33	0.31	7.05
WS_2	1130	0.24	0.26

Although all 2D materials in the tables above have a bandgap, the mobility is higher than in Si only in Hf-based materials, for which only scarce, on flake studies exist. Another observation is that for some 2D materials the effective mass is very anisotropic, fact that can be used in our advantage, as will be seen further regarding very small transistors that can avoid short channel effects.

The architecture of the Si FETs has evolved during time from top gate (Fig. 3.20(a)), to FinFET (Fig. 3.20(b)), and nanosheets FET (or multi-fin FinFET, Fig. 3.20(c)).

The challenge of FinFETs or nanosheets FETs is to reduce the leakage current when the transistor is in the *off* state, and in this respect the gate is covering entirely the channel (Ye et al. 2019). The term "nanosheet" has not the same signification for 2D transistors, where it means a flake obtained using a scotch tape from a 3D van der Waals material, whereas in Si FETs "nanosheet" refers to a periodic nanostructured channel with dimensions around 10 nm and even below, which is fabricated with advanced clean room technologies; the dielectric of the nanosheets channel is grown using ALD.

If we compare the state-of-the-art of the architecture of 2D FETs with the Si FETs architecture, we see that 2D FETs are in a primitive stage. The large majority of the results are based on 2D materials flakes with the substrate as a back gate. There are few results where 2D materials FETs are fabricated at the wafer scale. Among the few examples, we mention the batch fabrication of thousands of MoS_2 and WS_2 transistors at 4 in. wafer scale. The MoS_2 and WS_2 thin films are three-atom-thick semiconducting films. The growth method is Metal-Organic-Chemical

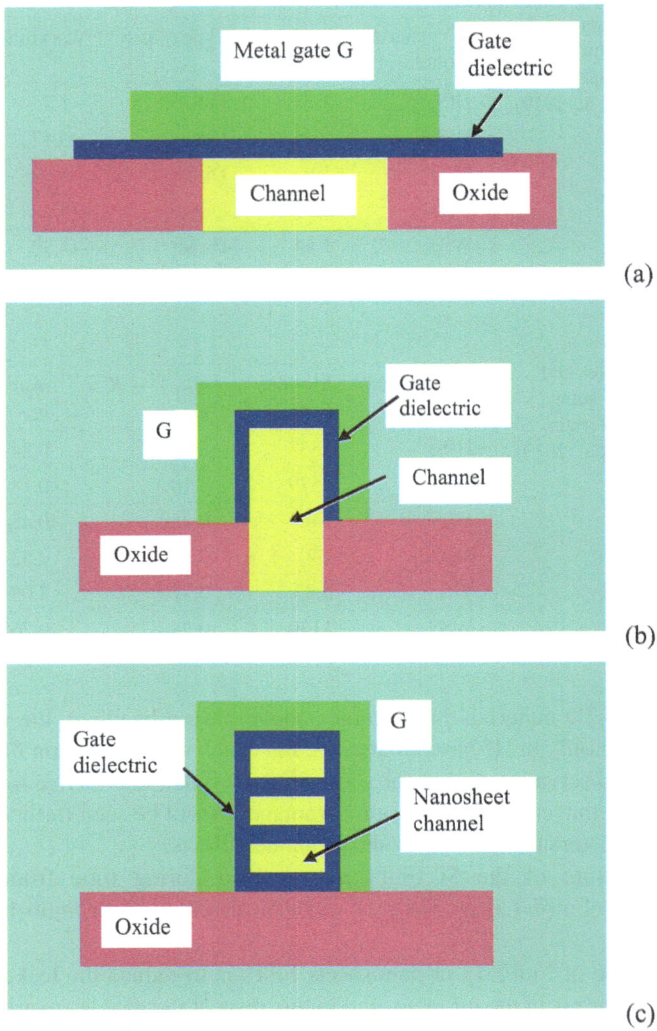

Fig. 3.20 Cross-sections through the channel of different Si FETs architectures: (a) planar FET, (b) FinFET, and (c) nanosheets FET

Vapor-Deposition (MOCVD) on a SiO_2 substrate, which in turn is thermally grown on a doped Si substrate working as a back gate (Kang et al. 2015).

The transistor configuration is a multi-electrode structure, such that the distance between drain and source could be varied in the range 1.6–34 μm (see Fig. 3.21). Electrical measurements have shown that the *on/off* ratio is 10^6, the carrier density is 7×10^{12} cm^{-2} for a back-gate voltage of 5 V, and that the mobility is 30 cm^2/V s.

There are some attempts to fabricate MoS_2 FETs in fin FET technology for demonstrating the integration of 2D materials in CMOS technology (see Fig. 3.22). In this

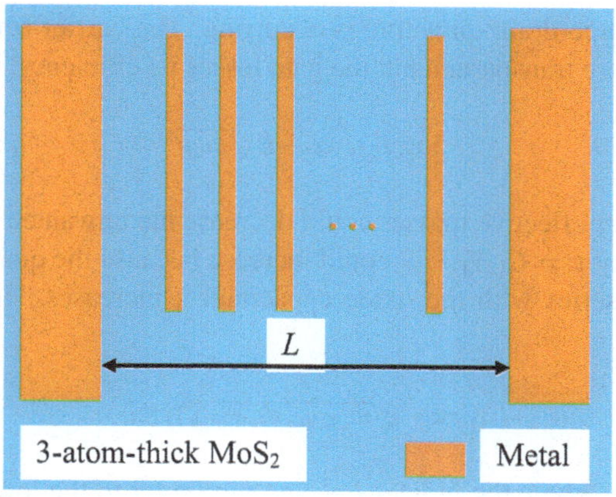

Fig. 3.21 The configuration of the electrode array on top of 3-atom-thick MoS_2

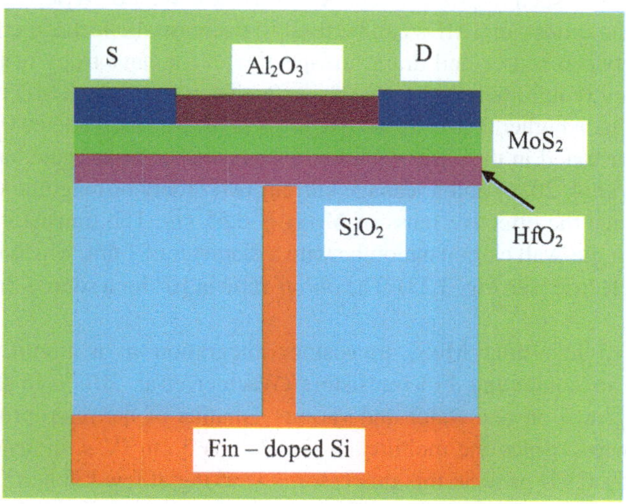

Fig. 3.22 10-nm gate length MoS_2 FET integrated with CMOS FinFET technology

respect, wafer-scale MoS_2 FET arrays were fabricated with a gate length of 10 nm, the MoS_2 layer being 0.7 nm thick (Pan et al. 2019). The *on/off* ratio is 10^6 for a drain voltage of 0.1 V, showing that the fin gate is depleting the channel with a high yield, and as a result the leakage current is 10 pA. The fabrication process consists of the following major steps: the fin Si is fabricated in CMOS technology, the MoS_2 is grown by CVD and transferred on the Si wafer, and then the source and drain contacts are patterned.

The 2D materials could extend the Moore's law to 1 nm due to the fact that many of them have very anisotropic effective masses (Ilatikhameneh et al. 2016). When the channel length L becomes short, heat problems appear and, moreover, a leakage current due to source-drain tunneling is occurring. The barrier between source and drain becomes more transparent and the gate losses its efficiency. We have then

$$\log I_{off} \propto -L\sqrt{m_{\text{eff}}} \tag{3.9}$$

meaning that larger effective masses could decrease the unwanted tunneling effects. However, the delay $\tau = C_G V_D / I_{on}$ could increase because the quantum capacitance C_Q, which is in series with the oxide capacitance, increases. However, if m_{eff} is anisotropic,

$$C_Q = e^2\sqrt{m_{\text{eff,low}} m_{\text{eff,high}}})/\pi\hbar^2 \tag{3.10}$$

is also decreasing, where the indices *low* and *high* refer to the low and high values, respectively, of the effective mass. So, the source-drain tunneling and the delay can be decreased simultaneously due to some 2D materials which have a highly anisotropic mass, such as phosphorene, HfS_2, $HfSe_2$, WS_2 or WSe_2. If the delay is not a concern, almost all TMDs can be used to decrease the leakage current due to tunneling between source and drain, irrespective of the anisotropy of the effective mass, the only requirement being that the effective mass of the TMD atomic-thick material be much higher than the effective mass of Si, considered to be 0.19 m_0. This fact is demonstrated in the case of MoS_2 FETs having a record gate length of only 1 nm (Desai et al. 2016) and a leakage current two orders of magnitude lower than a Si FET, since the effective mass of MoS_2 is 0.55 m_0. The transistor has as gate electrode a single-wall carbon nanotube with a diameter of 1 nm, which is embedded in a ZrO_2 dielectric (see Fig. 3.23). The *on/off* ratio is 10^6 for a source-drain distance of 1 μm.

The record regarding MoS_2 transistors integration in a circuit is a 1 bit-microprocessor containing 115 transistors (Wachter et al. 2017). In fact, only 18 FETs were grown on one wafer and various subunits of the microprocessor were bonded by wires. Since the mobility of the FET is 3 cm²/V s, indicating a lot of defects during CVD growth, this processor is working at few kilohertz, well below even the first Si microprocessors, which today look rudimentary.

Fig. 3.23 MoS_2 FET with CNT gate electrode

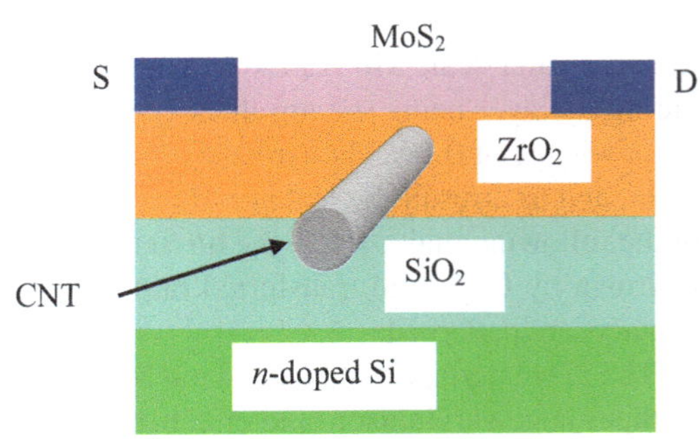

MoS_2 and other TMDs were used for high-frequency applications, such as switches (Kim et al. 2018), field effect transistors (Sanne et al. 2015; Liu et al. 2018) with a cutoff of 42 GHz (Cheng et al. 2014), and harvesting at few gigahertz (Zhang et al. 2018).

We have recently fabricated a microwave circuit based on a MoS_2 self-switching diode (SSD) (see Fig. 3.24) (Dragoman et al. 2020a) to detect radio waves. The SSD is a side-gate transistor with source and drain electrodes in shortcut. The entire detection circuit was fabricated at the wafer scale. The MoS_2 thin film (10-monolayers nominal thickness) was grown on a 4-in. Al_2O_3/high-resistivity silicon wafer by the CVD ALD deposition method. We have demonstrated experimentally that a microwave circuit based on a few-layers MoS_2 (see Fig. 3.25) is able to detect the audio spectrum from amplitude-modulated microwave signals in the band 0.9–10 GHz, i.e., in the frequency range mostly used by current wireless communications.

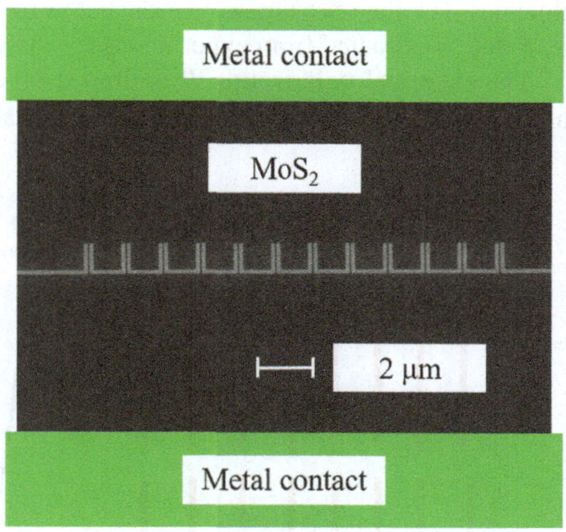

Fig. 3.24 The SSD MoS_2 diode

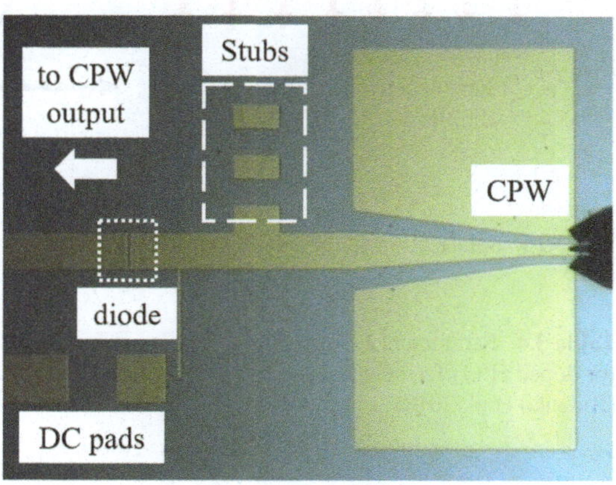

Fig. 3.25 MoS_2 microwave circuit containing MoS_2 SSD

3.1.3 Ferroelectric Transistors and Devices Based on Thin Film Ferroelectrics

The ferroelectrics were studied frequently in connection with transistors and devices, such as memories. More than 700 materials are ferroelectric, sharing in common the same physical property, i.e., they are insulators with ordered polarization (electrical dipoles) along two opposite directions that can be switched by an applied electric field (see Fig. 3.26).

The energy function describes the two states of polarization separated by a barrier, and using the Landau model (for a review see Wong and Salahuddin 2019) we have

$$U = \alpha P^2 + \beta P^4 + \gamma P^6 \tag{3.11}$$

Since in this section we refer only to ferroelectrics based on doped HfO_2 with thickness less than 5–6 nm, we present below in Table 3.4 the values of the parameters α, β and γ extracted from experiments for these materials.

The electrical polarization P is flipped by an applied DC field E greater than a coercive electric field E_C (which is a material-dependent parameter, typical for a

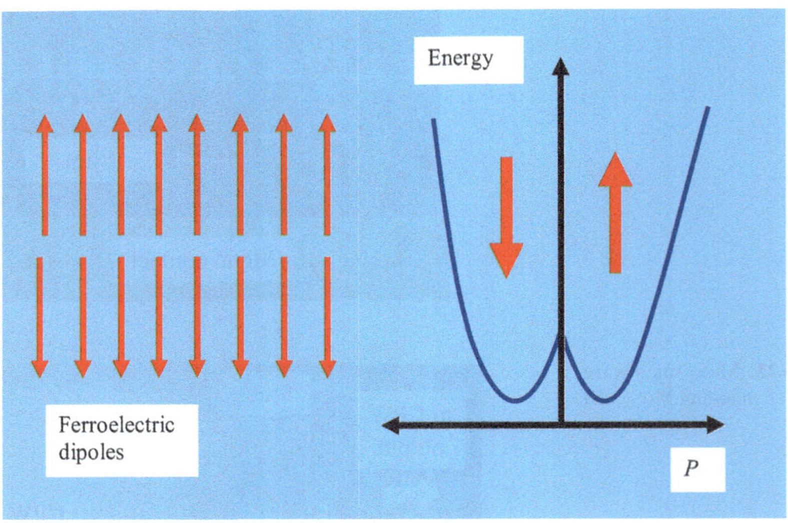

Fig. 3.26 Polarization states and energy in ferroelectrics

Table 3.4 Parameters α, β, γ for doped HfO_2 ferroelectrics (Bidenko et al. 2018)

Ferroelectric	α (m/F)	β ($m^5/F/C^2$)	γ ($m^9/F/C^4$)
HfO_2 doped with Zr (HfZrO)	-7×10^8	10^{12}	–
HfO_2 doped with Si	-7×10^8	-5.1×10^{10}	3.7×10^{12}

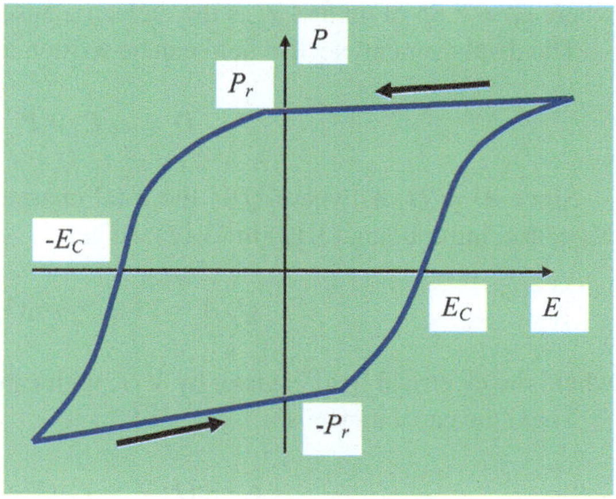

Fig. 3.27 *P–E* dependence in ferroelectrics

certain ferroelectric). Thus, a hysteretic behavior of the dependence *P–E* is produced (see Fig. 3.27), which is the imprint of any ferroelectric. The remanent polarization P_r is the polarization that remains in the ferroelectric after the electric field is reduced to zero, and is also a material-dependent parameter for a certain ferroelectric. An electric field higher than E_C (a poling electric field) disables the remanent polarization.

Ferroelectrics lose their properties beyond a temperature termed Curie temperature, losing also the piezoelectricity, which is a physical property typically associated to ferroelectrics. Basic physical properties of ferroelectrics can be found in Gonzalo and Jiménez (2007), Bain and Chand (2017) and many other textbooks.

The basic configuration to test the electrical properties of a ferroelectric device is a metal-ferroelectric-metal (MFM) capacitor (see Fig. 3.28), which is very similar to the well-known electronic devices termed MIM (metal-insulator-metal), the main difference between MFM and MIM being that the MFM capacitor displays always a hysteretic capacitance (*C*)–voltage (*V*) dependence.

The electrical polarization is given by

$$\boldsymbol{P} = \chi_e \varepsilon_0 \boldsymbol{E} \tag{3.12}$$

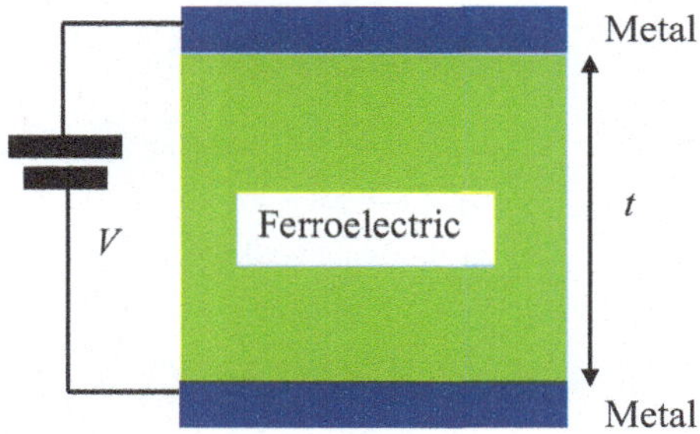

Fig. 3.28 The MFM capacitor

where $\varepsilon_0 = 8.85$ pF/m and χ_e is the dielectric susceptibility.

The displacement electric field can be written as

$$\boldsymbol{D} = \varepsilon_0 \boldsymbol{E} + \boldsymbol{P} \tag{3.13}$$

Since $D = Q/A$, where Q is the total charge and A is the area of the MFM capacitor, introducing (3.13) in (3.12) we get

$$Q/A = (1 + \chi_e)\varepsilon_0(V/t) \tag{3.14}$$

where the electrical field is given by V/t, with t the thickness of the ferroelectric.

Then, the capacitance is

$$C = Q/V = C_0(1 + \chi_e) \tag{3.15}$$

where $C_0 = \varepsilon_0 A/t$. Thus, there is a direct connection between polarization and capacitance, so that the basic physical parameters of the ferroelectric, such as the remanent polarization and the coercive field, can be found from capacitance measurements. In practice, there is special equipment able to measure $P(E)$ with high accuracy. Figure 3.29 is a schematic display of the typical C–V dependence of a MFM capacitor.

At the maximum value of the capacitance, the polarization of the dipole is reversed, and the coercive electric field is $E_C = V_C/t$. We observe that at zero voltage the capacitance is not zero, and this value provides information on the remanent polarization.

The MFS and MFIS (see Fig. 3.30) are other two types of ferroelectric capacitors very frequently used in many devices, where the letters S and I stand for semiconductor and insulator layer, respectively. These capacitors have a capacitance-voltage dependence that is different from that of the MFM, but the hysteretic nature

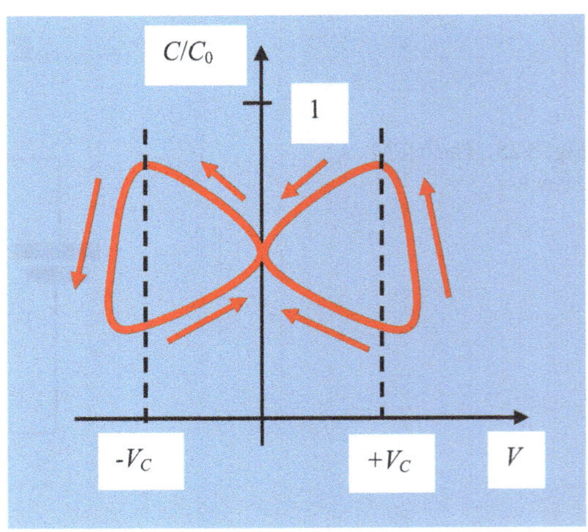

Fig. 3.29 Schematic representation of the MFM capacitance-voltage dependence

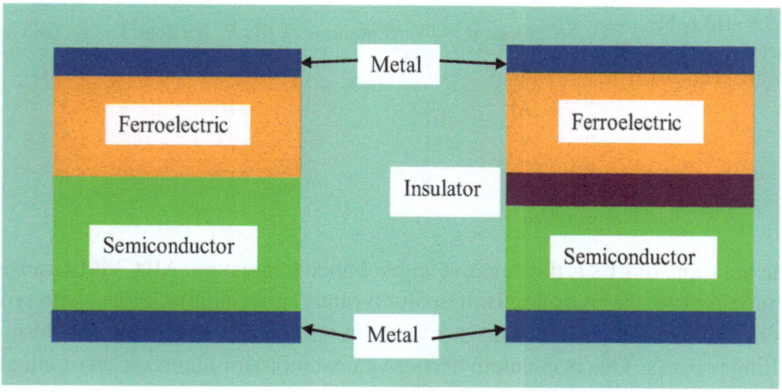

Fig. 3.30 Cross sections of MFS (left) and MFIS (right) ferroelectric capacitors

is preserved. In principle, the hysteresis has a clockwise rotation, but it happens frequently to be counter clockwise in MFS and MFIS due to many physical reasons.

There are many applications of ferroelectrics in nanoelectronics, which will be analyzed in this section:

- as negative capacitance transistors for sub-60 meV/decade subthreshold slope, i.e., for transistors biased at low voltages and with low leaky currents (Wong and Salahuddin 2019)
- as memories
- as piezoelectric materials, ferroelectrics being widespread in MEMS devices, for instance cantilevers and sensors, due mainly to their piezoelectric properties
- as microwave tunable devices, the relative permittivity ε_r being dependent on the applied DC voltage. Tunable phase shifters, filters, and antenna arrays change their main parameters, e.g., phase, resonant frequency or radiation pattern, by tuning the DC applied voltages.

There are many problems of ferroelectrics (FE) when applied to nanoelectronics. A general problem is that the most used FEs are perovskites, such as $SrBi_2Ta_9O_9$ (SBT) or $Pb(Zr_xTi_{1-x})O_3$, known as PZT, which are not compatible with CMOS technology. In addition these FEs have also very high dielectric permittivity and are not scalable, i.e., below a certain thickness they lose their ferroelectricity.

The ferroelectric phase of HfO_2 was discovered few years ago, by doping HfO_2—a widespread gate insulator in CMOS very large scale integrated circuits, such as microprocessors. The ferroelectric HfO_2 is few nanometers thick, generally below 6 nm, and the occurrence of ferroelectricity is due to an orthorhombic crystalline structure. The best known FE based on HfO_2 is HfZrO, which is HfO_2 having few nm thick and doped with Zr. There are many other dopants for FE HfO_2 such as Si, Al, and Y (Böscke et al. 2011; Müller et al. 2012; Schröder et al. 2013).

Depending on the dopant and its concentration, the main physical parameters of FEs based on HfO_2 are varying, as can be seen in Table 3.5.

Table 3.5 HfO_2 FE parameters for various doping (from Park et al. 2015)

Dopant	Concentration (%)	P_r (μC/cm^2)	E_c (MV/cm)
Si	4	24	0.8–1
Al	4.8	16	1.3
Y	5.2	24	1.2–1.5
Zr	50	17	1

The hot topic of FEs is the negative-capacitance transistors. All CMOS transistors are dominated by the so-called Boltzmann tyranny. In principle, Boltzmann tyranny refers to the fact that CMOS transistors cannot remove all heat generated during the switching process. This is the main barrier against more miniaturization of integrated circuits and the increase of their speed.

In the subthreshold swing (SS) we have

$$\begin{aligned}\mathrm{SS} &= \partial V_\mathrm{G}/\partial(\log I_\mathrm{D}) = (\partial V_\mathrm{G}/\partial\varphi_\mathrm{S})(\partial\varphi_\mathrm{S}/\partial(\log I_\mathrm{D}))\\ &= (\partial V_\mathrm{G}/\partial\varphi_\mathrm{S})(k_B T\ln 10/q)\end{aligned} \tag{3.16}$$

where I_D is the drain current, V_G the gate voltage, and φ_S the surface potential. The last term in (3.16), $k_B T\ln 10/q$, is 60 mV/decade at room temperature, and cannot be decreased. However, we can decrease $m = \partial V_\mathrm{G}/\partial\varphi_\mathrm{S}$, which is given by:

$$m = \partial V_\mathrm{G}/\partial\varphi_\mathrm{S} = 1 + C_\mathrm{G}/C_\mathrm{S} \tag{3.17}$$

where C_G is the gate capacitance, and C_S the semiconductor capacitance, both capacitances being connected in series.

In principle, $m \cong 1$, but C_G can be reduced if the gate dielectric is incorporating a FE with a negative capacitance. This concept will be explained below in more details. Finally, the total capacitance of the transistor is

$$1/C = (1/C_\mathrm{G} + 1/C_\mathrm{S}) > 0, \text{ while } m < 1 \tag{3.18}$$

The effect of the FE negative capacitance is depicted in Fig. 3.31.

The negative capacitance concept was introduced by (Salahuddin and Datta 2008) and is interpreted as positive feedback of the nonlinear capacitance-voltage dependence of the FE, stabilized by a series capacitance. A very intuitive approach of the negative capacitance is provided by Wong and Salahuddin (2019). Let's consider the FE FET represented in Fig. 3.32, where the dielectric insulator is replaced by a FE, which is also an insulator material.

The equivalent circuits of capacitors of the FE FET consists of two series capacitors: C_FE, which is the ferroelectric capacitor, and C_S, which is the equivalent capacitor of the semiconductor part of the FET. We have $V_\mathrm{G} = V_\mathrm{FE} + \varphi_\mathrm{S}$, where V_FE is the voltage across the FE. Now, the charges of the two capacitors are: $Q_\mathrm{S} = C_\mathrm{S}\varphi_\mathrm{S}$ and, respectively, $Q_\mathrm{FE} = \varepsilon_0 E_\mathrm{FE} + P$. Since $Q_\mathrm{FE} = Q_\mathrm{S}$, we have

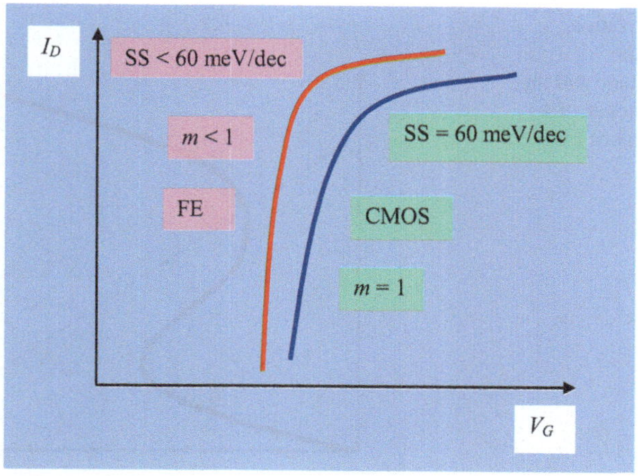

Fig. 3.31 Effect of FE negative capacitance on FET performances

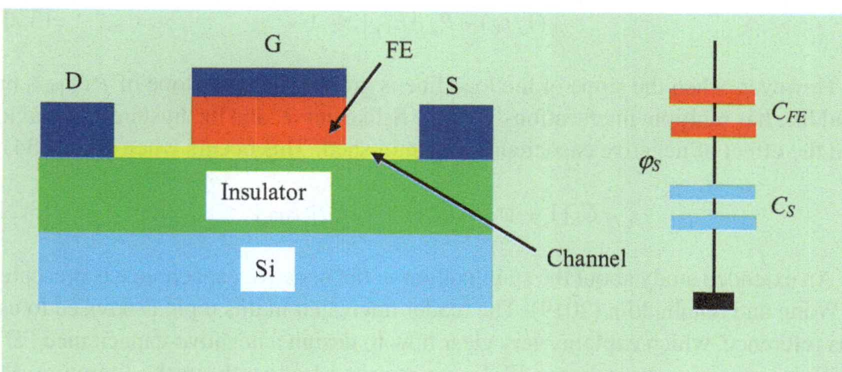

Fig. 3.32 FE FET and the equivalent circuit of the capacitors

$$P = C_S V_G - C_S[1 + (t_s/\varepsilon_r t_{FE})]V_{FE} \quad (3.19)$$

where t_s and ε_r are, respectively, the thickness and the relative permittivity of the semiconductor, modeled as a linear dielectric capacitor, and t_{FE} is the thickness of the ferroelectric.

Looking at (3.19), its lhs, i.e., P, is considered to be known from the Landau model or from measurements (i.e., the coefficients α, β, γ must be known, as follows from Table 3.4). The rhs of (3.19) is the load of the S-shaped dependence of P on V_{FE}, with the slope $C_S[1 + (t_s/\varepsilon_r t_{FE})]$; see Fig. 3.33.

When the load line is passing only through the negative slope of $P(V_{FE})$, we have a solution for the negative capacitance in the region $\partial P/\partial E_{FE}$, where $E_{FE} = V_{FE}/t_{FE}$. Thus, the negative capacitance is occurring when

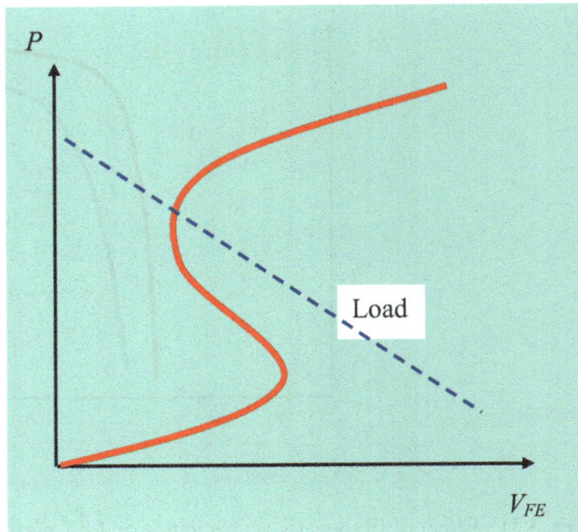

Fig. 3.33 Polarization dependence on the ferroelectric voltage and the load for the existence of a negative-capacitance FET

$$(1/\varepsilon_0)|\partial P/\partial E_{\mathrm{FE}}| \gg 1 \tag{3.20}$$

However, when the slope of the load line is greater than the slope of $P(V_{\mathrm{FE}})$, the load line has multiple intersections with the S-like curve, and the hysteretic behavior and the effect of negative capacitance is diminished. This occurs when (Fig. 3.34)

$$C_{\mathrm{S}}[1 + (t_s/\varepsilon_{\mathrm{r}} t_{\mathrm{FE}})] > |\partial P/\partial E_{\mathrm{FE}}| \tag{3.21}$$

An extended study about the stabilization of the negative capacitance is presented in Wong and Salahuddin (2019). The reader interested in this topic is advised to use this reference, which explains very clear how to design a negative-capacitance FET.

Today, negative-capacitance FETs are reported frequently in the literature. We refer here only at FETs based on HfO_2-based ferroelectrics, which are CMOS compatible. The combination of 2D materials as channel and HfZrO as ferroelectric

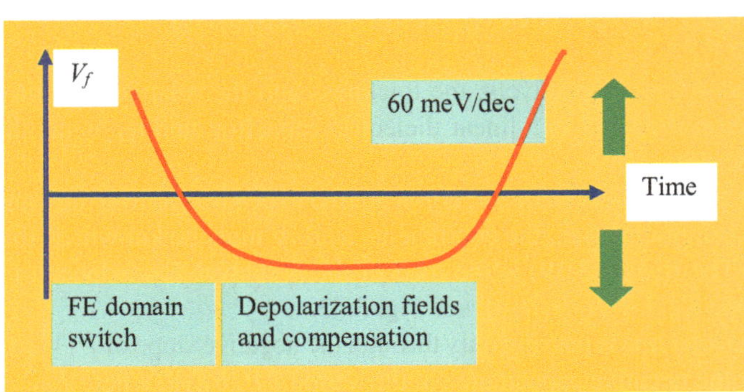

Fig. 3.34 The time domain behavior of the voltage on MFIS

constitutes the most advanced to date atomic scale FETs. Some results are listed in Table 3.6, including FinFETs—the state-of-the-art of CMOS technology.

It is interesting to investigate what happens in time domain regarding the negative capacitance effects in MFIS structures (Sharma et al. 2018). In this case, the insulator, such as SiO_2, is intercalated between Si and the FE, and an additional capacitance C_I must be put in series with C_{FE} and C_S in the equivalent series circuit modeling the MFIS configuration. Note that the voltage that drops on all these series capacitors is denoted by V_f. If the DC voltage supply V_S, in series with the resistor R_S, is exciting the MFIS structure and is larger than the coercive voltage, we have $\mathrm{d}P/\mathrm{d}V_f > 0$ if the resistor is absent. However, the inequality changes sign if the resistor is present, indicating a negative capacitance effect. The negative capacitance effect is due to depolarization fields, producing high polarized charges, which are compensated after a certain time by the charges pumped by the DC voltage source. Figure 3.35 represents

Table 3.6 Boltzmann tyranny abolished

Transistor type	Gate length (nm)	FE	I_D–V_G hysteresis	SS at room temperature (mV/dec)	References
FinFET (state-of-the-art CMOS)	14	Si-doped HfZrO, 4-8 nm thick	No	54	Krivokapic et al. (2017)
FET, MoS_2 monolayer channel (back-gate)	2000	HfZrO	No	20	Si et al. (2018)
FET, MoS_2 channel (back-gate)	1000	HfZrO	Yes	6	McGuire et al. (2017)

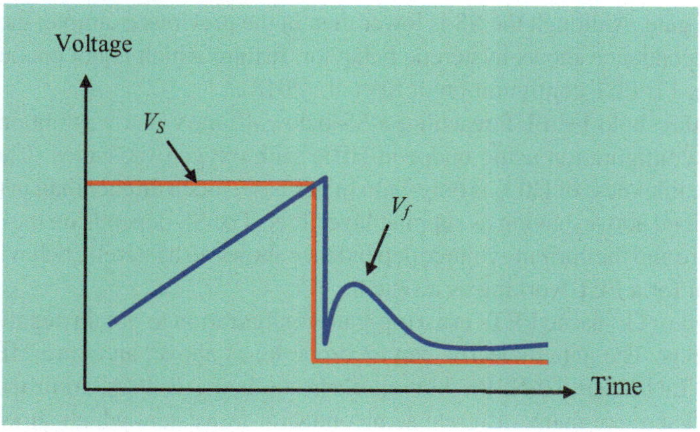

Fig. 3.35 The imprint in time of the negative capacitance

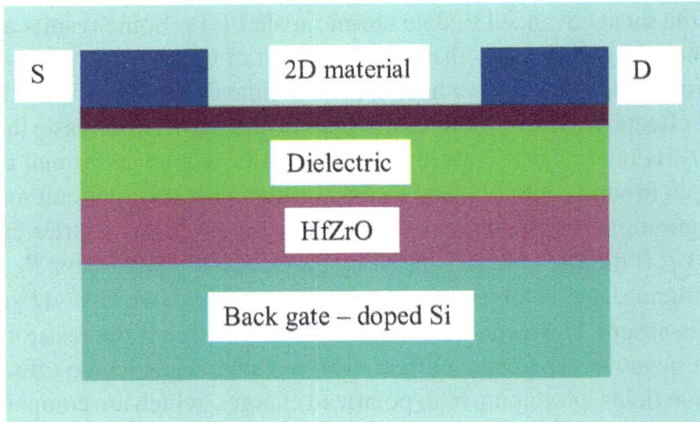

Fig. 3.36 Back-gate configuration of the negative capacitance FE FET

schematically the time dependence of V_f. The hump in the curve of V_f is the imprint of the negative capacitance, the shape of the hump depending on the values of the resistance R_S in series with the DC source.

Many negative capacitance transistors were reported with SS below 60 meV/dec. In the case of 2D materials, the basic configuration is a back-gate transistor, such as that depicted in Fig. 3.36

In (Si 2018), MoS_2 was used as a 2D material, with the channel length of 1 μm, width of 1.56 μm and thickness of 8.3 nm, obtaining SS around 20 meV/dec depending on the growth temperature. The dielectric with the thickness of few nanometers is the capping layer, with the role of reducing the leakage current and for capacitance matching. In (McGuire et al. 2017) a different strategy was adopted, by integrating a MFM capacitor in a 2D transistor, and thus reaching a SS of 8 meV/dec (see Fig. 3.37). In this configuration there are two electrodes: V_{int}, used to alleviate the interfacial problems and to control the gain of the transistor, and V_{NC}, which plays the role of the gate. Although the SS is lower than in the previous example, the current–voltage dependence shows hysteretic behavior, feature which is not encountered in the simpler FE FET configuration of Si et al. (2018).

A sub-threshold FE FET reaching a SS below 40 meV/dec was obtained in the back-gate configuration using undoped HfO_2 with oxygen vacancies (Cheng et al. 2019). The thickness of HfO_2 is only 4 nm in this case, much thinner than in the examples described above, having as capping layer TaN. The SS depends on the annealing temperature and the current–voltage dependence shows no hysteretic behavior, which is essential for a FET working as a switch.

Si-based or Ge-based FETs use a top-gate configuration to obtain negative capacitance effects. The gate dielectric can incorporate, as above, an entire MFS capacitance, as TaN/HfZrO/TaN, the bottom electrode being isolated from the channel by HfO_2. There are many different semiconductor transistors which show negative capacitance and SS below 60 meV/dec. A recent review (Alam et al. 2019) discusses

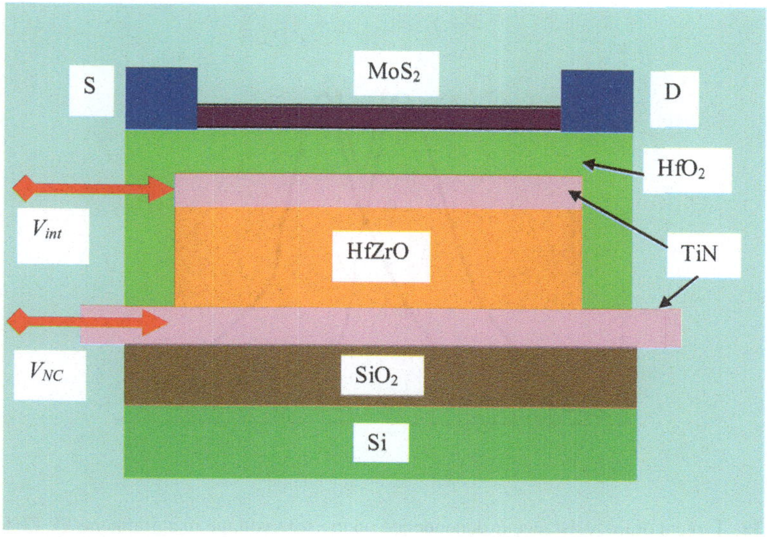

Fig. 3.37 The MFM capacitor FE FET

the SS for different semiconductor FETs using different ferroelectric types, not only HfZrO.

The ultimate negative-capacitance FET would involve one monolayer channel and one monolayer FE, both being van der Waals materials. There are 2D materials that exhibit ferroelectricity, such as $CuInP_2S_6$ (CIPS) or $InSe_3$ (see Cui et al. 2018). The first attempt to achieve a negative-capacitance FET based on MoS_2 and CIPS regarded a top-gate transistor showing SS of 28 meV/dec (Wang et al. 2019b) (see Fig. 3.38). Although in this first attempt, neither the channel nor the FE have a thickness of one monolayer and the channel length was large, of 5.7 μm, the path to achieve an ultimate negative-capacitance transistor was set.

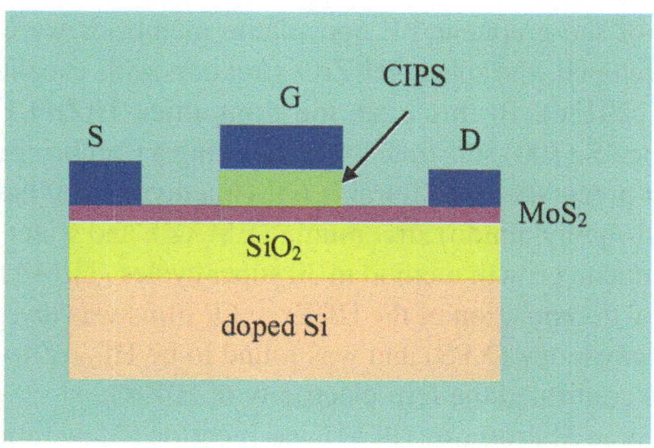

Fig. 3.38 FE FET with 2D material channel and FE

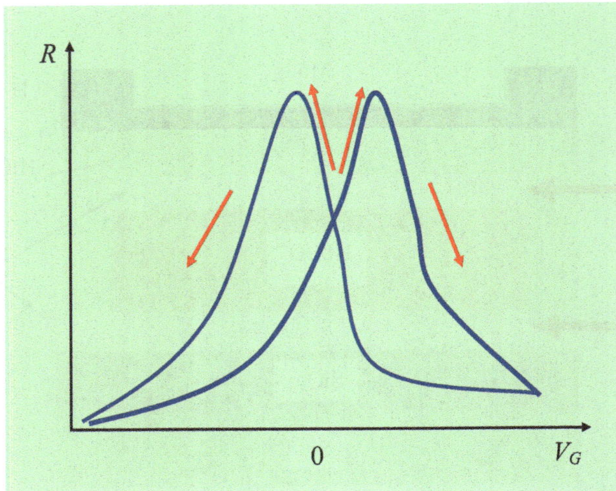

Fig. 3.39 Typical drain resistance *R* dependence on the gate voltage in a graphene/FE FET

Graphene was not used for negative capacitance transistors, but graphene/FE was used for memory applications and other FET functionalities, which will be described in the following. In principle, a non-volatile graphene memory is a top-gate FET configuration, where the gate dielectric is replaced by FE. The accumulation and depletion of carriers in the channel is controlled by the gate voltage, so that the polarization of the FE is inducing a hysteretic dependence of the drain voltage or drain resistance (*R*) on the gate voltage (see Fig. 3.39). When the gate voltage is set to the threshold value $\pm V_C$, where $V_C = E_C t$, with E_C the coercive electric field and *t* the thickness of the ferroelectric, the carriers are depleted and *R* is becoming very low. So, there are two distinct states corresponding to accumulation and depletion, respectively. There are excellent reviews regarding graphene/FE non-volatile memories (Jie and Hao 2014; Zhou and Chai 2017). These memories are non-volatile resistive memories since the memory effect involves a resistance and not a capacitor.

Because none of the graphene/FE nonvolatile memories were CMOS compatible, we have developed a graphene/HfZrO structure with excellent performances (Dragoman et al. 2018c). In this case, the 6-nm-thick HfZrO FE was grown at 250 °C on a *p*-doped Si(100) substrate by ALD, using a Cambridge NanoTech F200 reactor. The ALD precursors are: Tetrakis (ethylmethylamido) hafnium (TEMAH) and Tetrakis (ethyl-methylamido) zirconium (TEMAZ), and water. The FE film was deposited by a laminate growth method in 30 super cycles: TEMAH-H_2O-TEMAZ-H_2O. The chemical composition of the $Hf_xZr_{1-x}O_y$ films was investigated by X-ray photoelectron spectroscopy (XPS) and was found to be $Hf_{0.45}Zr_{0.55}O_{1.76}$. XRD and PFM studies were confirmed the ferroelectricity of HfZrO.

Piezoelectric Force Microscopy (PFM) was mentioned few times before. PFM is obtained using the NT-MDT Solver Pro P-47 AFM. Measurements are made in the piezoresponse force mode with a low-stiffness conductive probe (platinum-coated tip with a spring constant of 0.003–0.13 N/m (Tips Nano, CSG01/Pt)). A bias is applied between the conductive tip and the FE. Height and phase images were obtained by using 256 × 256 (x, y) positions (pixels). The ferroelectric behavior is evidenced in PFM measurements by the well-defined dark and bright contrast areas created as a function of the applied bias, i.e., by different polarization orientations.

In the PFM, the conductive probe is brought into contact with the sample, i.e., the AFM is working in the direct contact mode. The piezoelectric activity is induced with an AC voltage applied between the tip and the sample. The AC excitation, which has a frequency around 100 kHz in our case, is producing periodic expansion or contraction of the sample. The changes of the sample dimensions are detected by the cantilever deflection. In this technique, the amplitude (PFM amplitude) and phase (PFM phase) of the vertical tip displacement are recorded by a lock-in amplifier. Topography, amplitude and phase images can be measured separately for full PFM characterization. For example, Fig. 3.40 represents the topography and PFM phase images of the HfZrO thin film.

The PFM phase curve is recorded in a single point, with the tip in contact with the surface, by varying the amplitude of the AC signal (see Fig. 3.41). The hysteretic behavior and the phase change of 180 degrees is a direct proof of the ferroelectricity of the HfZrO probe.

After HfZrO was deposited and characterized, graphene was transferred on half of the 4-inch wafer, the other half being reserved to fabricate a MFS capacitor, as illustrated in Fig. 3.42(a). We have fabricated 88 metal-HfZrO-semiconductor (MFS) capacitors, and the same amount of metal-graphene-HfZrO-semiconductor-metal (GMFS) capacitors (Fig. 3.42(b)). The Cr (5 nm)/Au (100 nm) top electrodes with dimensions of 150 μm × 150 μm were deposited using an e-beam evaporation system (see Fig. 3.43).

We have displayed the C–V dependences in Fig. 3.44 for different sweeping ranges, at a frequency of 100 kHz for both types of capacitors. All C–V characteristics

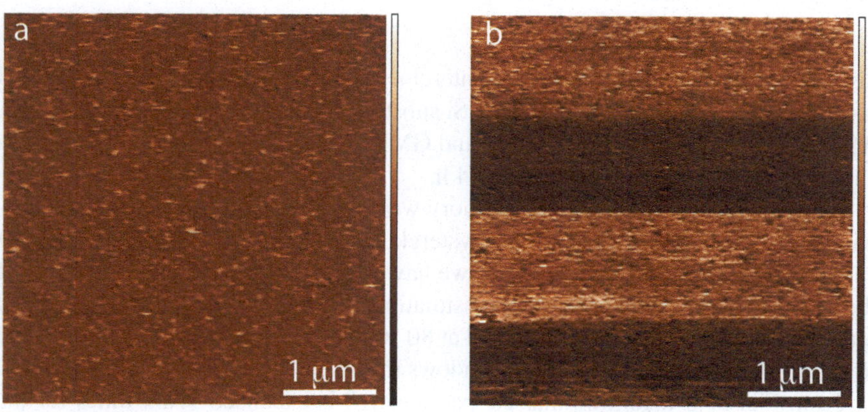

Fig. 3.40 (a) Topography and (b) PFM phase images of the HfZrO thin film (Dragoman et al. 2018c)

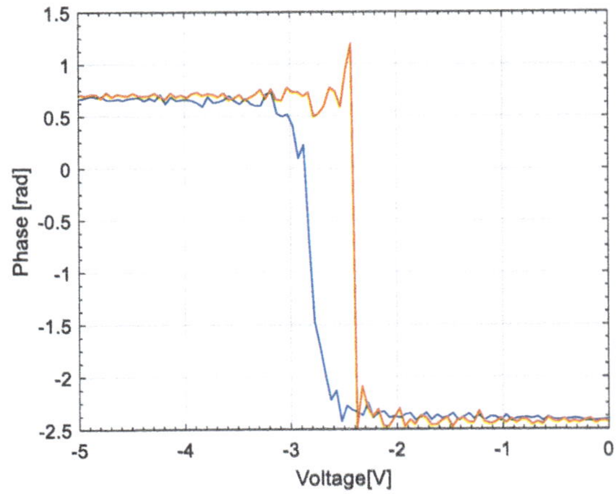

Fig. 3.41 PFM phase curve recorded in a single point as a function of the amplitude of the AC signal (Dragoman et al. 2018c)

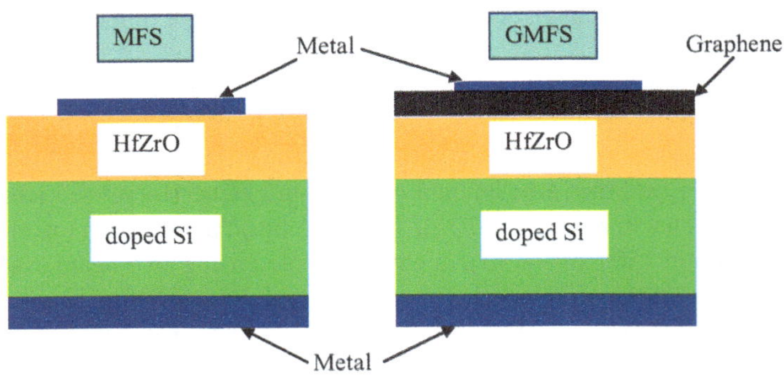

Fig. 3.42 MFS (left) and GMFS (right) capacitor configurations

display a hysteretic behavior with counterclockwise direction due to the interfacial layer between the ferroelectric and the Si substrate, formed during ALD deposition. After the measurements of both MFS and GMFS capacitors, we have performed an annealing at 80 °C in vacuum during 24 h.

It is interesting to compute the memory window in both situations. The memory window is the width in volts of the hysteretic behavior of $C(V)$ measured at half value of the capacitance. In Fig. 3.45, we have represented the memory window of the MFS and GMFS capacitors in two situations: for the samples as fabricated and after an annealing treatment performed at 80 °C in vacuum during 24 h. We defined R as the ratio between the memory windows in GMFS and MFS.

From Fig. 3.45 it follows that the parameter R is enhanced 17.35 times for the sweeping range (−4,4) V after annealing, when the graphene monolayer is placed over the FE. Why is the memory window of GMFS much larger than that of MFS?

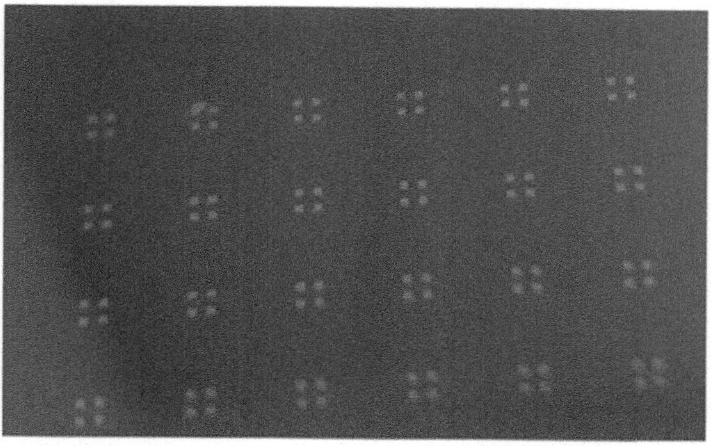

Fig. 3.43 Top electrodes of MFS and GMFS capacitors

The answer is that part of the applied voltage drops over graphene, enhancing the polarization P of the ferroelectric layer.

The graphene/FE structures are of interest in very innovative devices, such as reconfigurable transistors (Dragoman et al. 2020b). These graphene/FE FETs are either horizontal or vertical transistors, depending on the applied gate polarity. Under one gate polarity, the FET is working as a normal graphene FET, having a horizontal transport of carriers between the drain and the grounded source, the drain current being modulated by a back gate, whereas by changing the gate polarity, the transport becomes vertical, between the drain and the back gate.

For a better understanding of the operation principle of the device, consider first that the gate voltage V_G applied on the p-Si substrate is negative. Then, the device, which is represented in Fig. 3.46, behaves as a typical graphene FET, the drain current I_D flowing in the horizontal direction, between source and drain contacts. In this case, graphene is playing the role of the transistor channel and HfZrO the role of the gate dielectric. Thus, source and drain are metallic contacts with the typical role attributed in any FET, while ϕ_m denotes the workfunction of the metal M. On the other hand, when the back gate becomes positive, the direction of current flow is changing to vertical, since the charge carriers traversing the device originate mainly in p-Si. Thus, p-Si becomes the effective S electrode, as it is denoted by the parentheses in Fig. 3.46(b), left. The charges produced by the back gate must overcome the thin 6 nm HfZrO layer to reach the D electrode. However, their behavior is not that of carriers in a standard diode since the third electrode is modulating the current between p-Si and D, and hence the former S electrode becomes a gate electrode (role indicated in the parenthesis) in the case of the vertical transport.

The HfZrO layer is polarized, being located between the doped Si layer and graphene. Thus, a weak electric field is applied on the ferroelectric material as the drain voltage V_D raises towards positive values. The equivalent electric field on HfZrO decreases (the dashed red lines in the figures at right in Fig. 3.46, as response

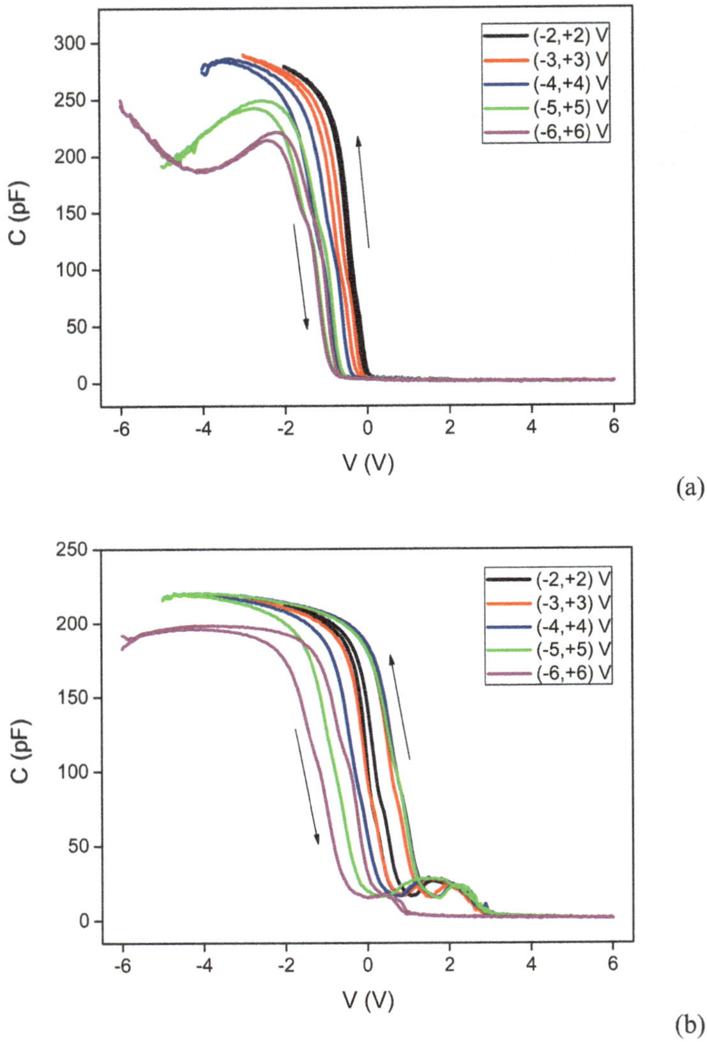

Fig. 3.44 Capacitance of (a) MFS and (b) GMFS capacitors versus voltage for different voltage sweeps, at a frequency of 100 kHz, both before thermal annealing (Dragoman et al. 2018c)

to the shift of the Fermi level in graphene, represented by the arrow), and thus the barrier height seen by charge carriers from *p*-Si increases. As a result, the current collected by the D electrode decreases. This constitutes the modulation effect in the vertical transistor, where S is playing the role of a gate. In Fig. 3.47(a), a SEM photo of the graphene monolayer channel, cut in a bowtie shape, is presented, while Fig. 3.47(b) shows the optical image of the chip of graphene/HfZrO FETs (right) and the SEM of a single FET (left).

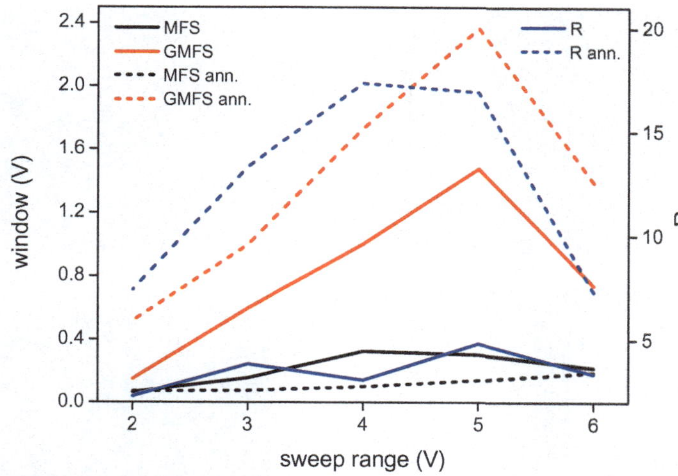

Fig. 3.45 Memory windows of MFS and GMFS before (solid line) and after (dotted line) annealing (Dragoman et al. 2018c)

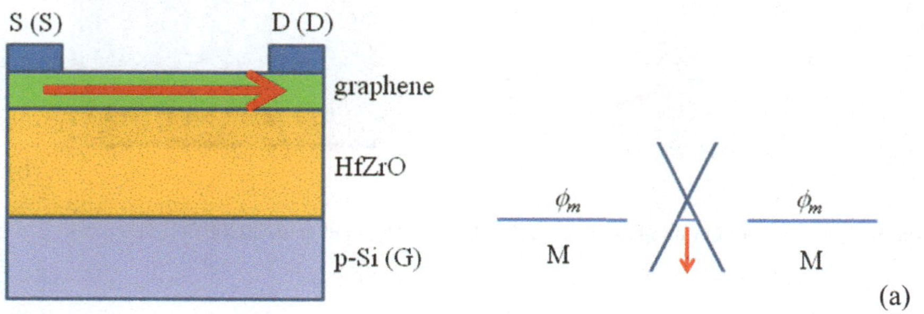

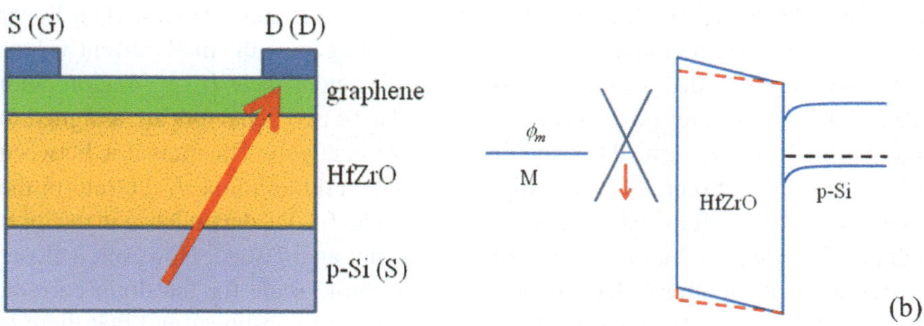

Fig. 3.46 Device configuration and band diagrams when no bias is applied for the (a) horizontal, and (b) vertical transport regimes (Dragoman et al. 2020b)

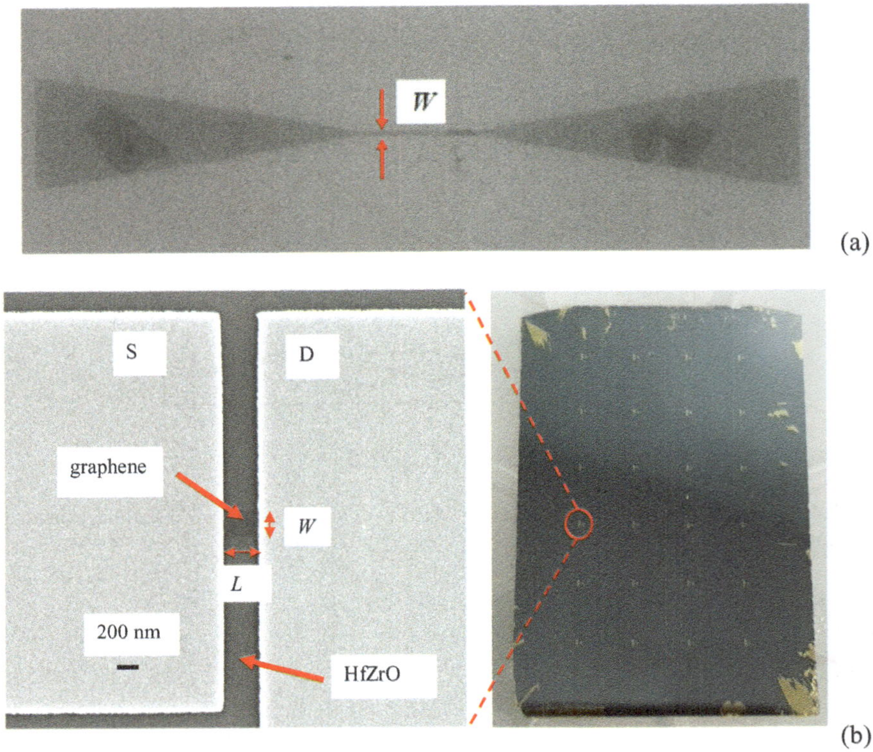

Fig. 3.47 The reconfigurable graphene/HfZrO FET (Dragoman et al. 2020b)

The drain current I_D versus gate voltage V_G dependence for various drain voltage values V_D indicated in the legend for graphene/HfZrO FETs with Ti–Au contacts are displayed in Fig. 3.48(a). We distinguish two main dependences: (i) a linear I_D–V_G dependence region at rather large drain voltages, with small current values at negative V_G (detailed also in the inset), and (ii) a non-linear I_D–V_G region, with large current values, for positive gate voltages. These two behaviors are assigned to (i) horizontal and (ii) vertical transport regimes, respectively. The transition between these two transport regimes has as distinct imprint a hump in the I_D current in the inset of Fig. 3.48(a) for small positive V_G values. The I_D–V_G dependence in the inset seems to be linear, so that the graphene FET has not an *off* state. However, a closer examination of the above dependences in logarithmic scale for the drain current, shown in Fig. 3.48(b) reveals that the dependences are non-linear and that there is three-orders-of-magnitude variation in the drain current when the gate voltage is swept from near 0 to −20 V at low drain voltages.

These non-linear dependences are assigned typically to a FET with a bandgap channel and not to a graphene FET, where there is no *off* state because there is no bandgap. The determined bandgap from the *on*/*off* ratio is 0.18 eV. Atomistic simulations show a bandgap induced in graphene by HfZrO of about 0.24 eV (Nemnes

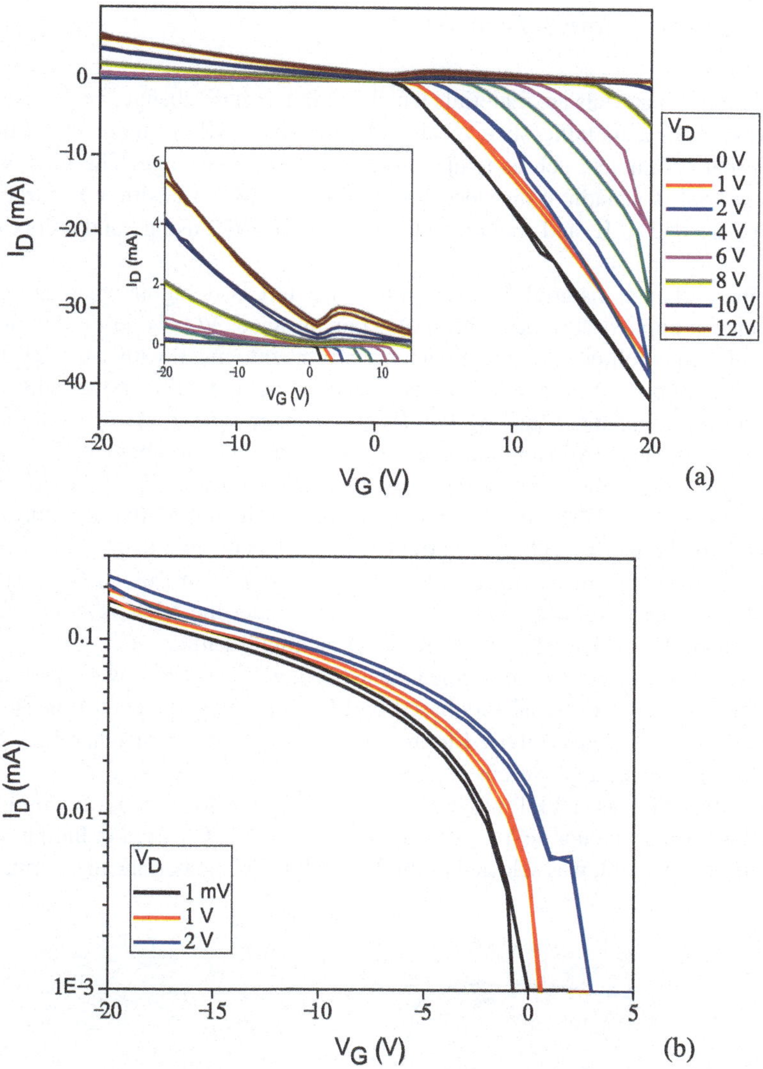

Fig. 3.48 Drain current versus gate voltage dependence of the reconfigurable graphene/HfZrO transistor (Dragoman et al. 2020b)

et al. 2019), mainly due to orbital hybridization and locally deformed graphene structure. This is a very important result since the other two materials able to induce a bandgap in graphene: SiC and h-BN, are not CMOS compatible, while HfZrO is able to be grown on vey large Si wafers.

3.1.4 Phase-Change Transistors

Phase-change materials are materials which exhibit a phase change, i.e., a reversible transition from a metallic state to an insulator state (MIT), under the action of an external parameter such as temperature, pressure, or voltage. The most known MIT material is vanadium dioxide, which shows an MIT transition at a transition temperature of 340 K, and was predicted by Mott in 1949 using band theory (Mott 1949).

In phase-change materials, the crystalline structure is changing when an external parameter is applied at a clear threshold. For instance, VO_2 is an oxide, an insulator with a monoclinic structure, which transforms into metal with rutile crystalline structure at 340 K. As a result, the resistance of the VO_2 film shows a hysteretic behavior, the resistance changing five orders of magnitude at $T_t = 340\,\text{K}$ (67 °C) (see Fig. 3.49). The reversible metal-insulator transition takes about 1–2 ps.

A recent review about phase transition mechanisms in vanadium dioxide can be found in Shao et al. (2018). It is interesting to note that in a MIM-like structure where the insulator layer I is VO_2, the current-voltage dependence displays a S-shaped negative differential resistance (NDR) region, as shown in Fig. 3.50 (Pergament et al. 2015). The threshold voltage necessary to trigger the NDR is $\approx (T - T_t)^{0.5}$ and is around few volts, while the current is reaching 0.1 mA.

Phase-change transistors are dedicated also to low SS for low power applications. There are two types of transistors based on MIT materials: the Mott transistor and the phase transistor, also referred to as negative differential resistance FET (NDR FET) for low power applications.

The Mott FET is a FET having as channel a thin MIT material. Initially, a perovskite-structure cuprate $Y_{1-x}Pr_xBa_2Cu_3O7\text{-}\delta$ (YPBCO), used in high temperature superconductors, was selected as the Mott MIT. This material can be epitaxially

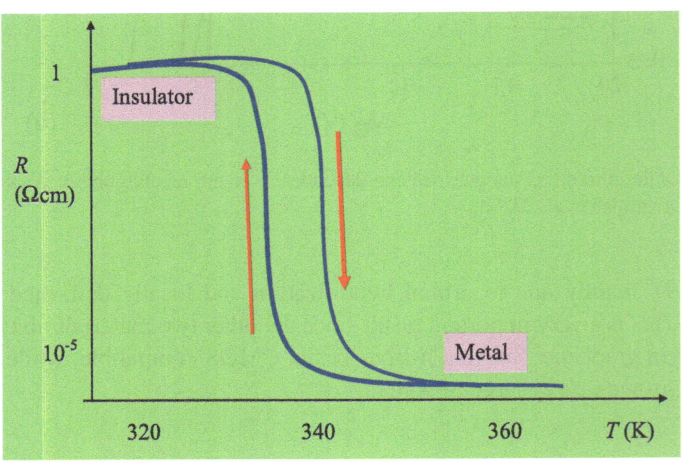

Fig. 3.49 MIT transition in VO_2 as temperature is varied

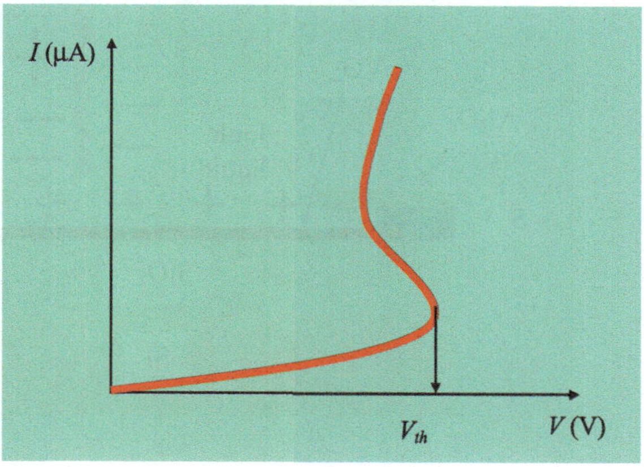

Fig. 3.50 NDR in VO_2 MIM-like structures

grown on the strontium titanate (STO) crystal. YPBCO displays a Mott transition from insulating to a metallic state by injecting in cuprate charges from a back gate. The conducting channel of *p*-type is formed at the interface cuprate/STO, and has a thickness of 1 nm (see Fig. 3.51). In this Mott FET, the drain current in the *on* state is reaching 40 μA at a drain voltage of 2.5 V and a gate voltage of −12 V, and is much reduced in the *off* state, for gate voltages below −2 V. This initial Mott FET was not further developed because its constitutive materials are very peculiar.

The present developments are focused on VO_2 Mott FET. In this case, the solid state oxide gates, such as HfO_2, do not confer enough drain conductance modulation (e.g. only 6% modulation was obtained by Sengupta et al. 2011), while a hysteretic behavior of the conductance dependence on gate voltage (i.e., a resistive memory effect) was observed. An improvement of the electrostatic gate modulation was recently achieved using liquid ionic gates (Ji et al. 2012). Here, the conductance is varying from 0.06 to 0.12 μS when the gate voltage is varied in the range −0.2 to 1.4 V, and the resistive memory effect is evidenced at room temperature. Such a transistor is presented in Fig. 3.52. By using such a ionic gate, as well as

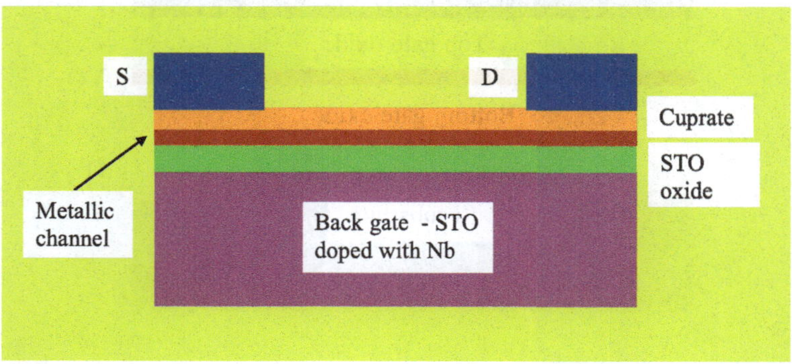

Fig. 3.51 The Mott FET

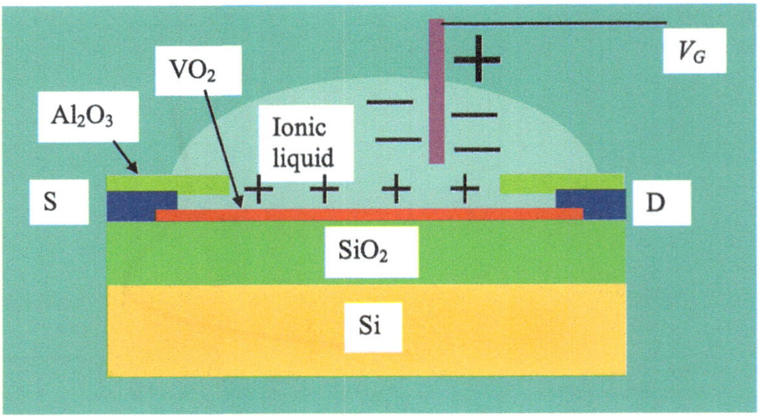

Fig. 3.52 Mott FET with a liquid gate

a strained VO_2 film of 10 nm thickness, a Mott FET was fabricated (Nakano et al. 2012). The electrostatic charge on the surface of VO_2, via the liquid gate, is driving the device into a 3D metallic state at a gate voltage of 1 V and the conductance is modulated with two-three orders of magnitude. The Al_2O_3 layer, with a thickness of 100 nm, has a protection role. Since the Mott FET displays a non-volatile resistive memory, it is envisaged to be used in future memory devices in a MIM configuration, or in a capacitive memory configuration, as that illustrated in Fig. 3.53 (Zhou and Ramanathan 2015).

The phase transistor (or hyper FET) is rather different compared to the Mott FET. In a phase FET, VO_2 is utilized as threshold selector (TS) by connecting a VO_2 thin film in series with the source of the FET (see Fig. 3.54) (Shukla et al. 2015). When the

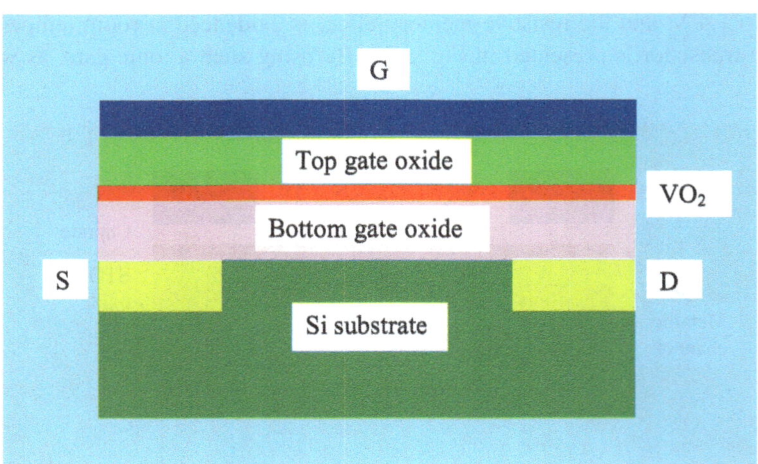

Fig. 3.53 Capacitive Mott memory

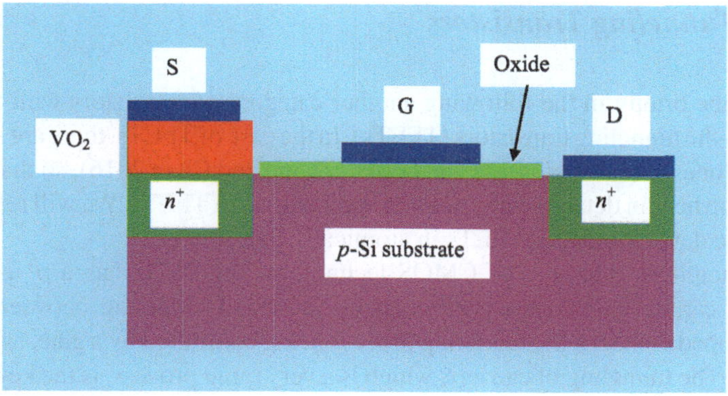

Fig. 3.54 The phase FET

gate voltage is zero, the TS is in the insulator state. By increasing the gate voltage, the channel resistance decreases and the drain current increases up to a threshold value $I_{D,TM}$, when the phase transition in VO_2 towards a metal state is taking place. As a result of MIT, the drain current suddenly jumps, and NDR is formed during this process. When the gate voltage decreases again, the channel resistance is increasing, the drain current is decreasing up to a threshold value $I_{D,TI}$ when the phase transition from the metallic into the insulating state takes place again, after which the drain current is decreasing rapidly. Due to the difference between the $I_{D,TM}$ and $I_{D,TI}$ values, the dependence of the drain current on the gate voltage has a hysteretic behavior (see Fig. 3.55).

The hysteretic nature of the phase FET recommends this transistor for resistive memory application, since the *on*/*off* ratio in the drain current is about two orders of magnitude. However, even if SS is lower than 60 meV/decade, the hysteretic behavior is an adverse factor for the utilization of the phase FET in logic applications. Unlike in negative capacitance transistors, where the hysteretic behavior inherent to ferroelectrics is much diminished via matching with a capacitance, in the case of phase FET the hysteric behavior is unavoidable due to the MIT transition.

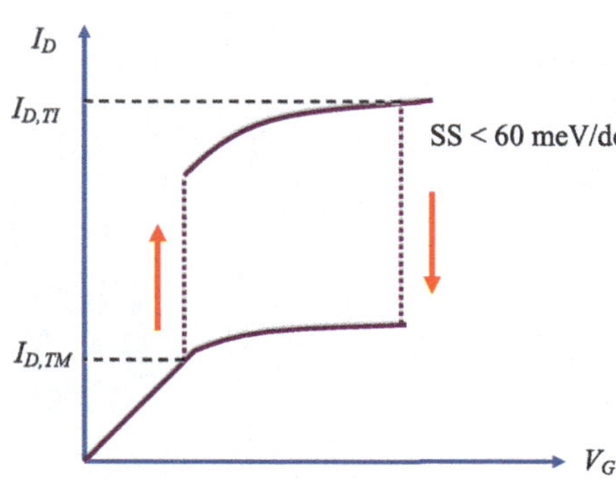

Fig. 3.55 Drain current versus gate voltage dependence in a phase FET

3.1.5 Tunneling Transistors

We describe briefly in the following another category of transistors with SS below 60 meV—the tunneling transistors (TFETs). In the case of TFETs, there are important reviews (Ionescu and Riel 2011) and books (Zhang and Chan 2016), so that, we will not develop here in detail the physics and applications of TFETs. We will refer briefly only to the development of TFETs in atomically thin materials.

In principle, in the case of CMOS technology, the TFET has a p^+-i-n^+ structure. The carrier injection is performed by intraband tunneling between the two heavily doped contacts, the tunneling process being controlled by a gate, as shown in Fig. 3.56. The tunneling of carriers, which is a very rapid process, is the key physical process of abrupt SS. The conduction in TFET is generally ambipolar but, by using various doping profiles or heterostructures, *n* or *p*-type TFETs can be obtained.

The discovery of 2D materials has enriched the TFETs configurations. For example, a graphene/TMD/graphene vertical structure, combining tunneling over a WS_2 barrier with thermionic transport, has produced an *on/off* ratio of 10^6 (Georgiou et al. 2012). The two graphene layers play the role of S and D electrodes, while the gate voltage, which controls the Fermi level in the structure, is applied between the doped Si substrate and the bottom graphene layer. The high *on/off* ratio is due to tunneling of the few-atomic-layer WS_2 barrier and to thermionic transport over it. The structure is that of a vertical back-gate FET, illustrated in Fig. 3.57.

There are also horizontal 2D tunneling transistors, as described in the recent review of Lv et al. (2019), which have an improved *on* state compared to vertical configurations. This occurs especially when 2D materials with narrow bandgaps are used, as for instance GNRs. For example, a GNR with a width of 10 nm has a bandgap of 40–50 meV. Since a horizontal TFET implies a *p-i-n* configuration and doping of GNRs are inefficient, three gates are used for electrostatic doping in this case (see Fig. 3.58) (Hammam et al. 2018). The gates G1 and G3 *p*-dope and, respectively, *n*-dope the GNR, and play the role of source and drain electrodes, while G2 is modulating the tunneling in the *on* or *off* state. The obtained SS is about 47 meV/dec up to 40 K, above which it is dramatically degraded. The saturation current is 8.5 μA/μm.

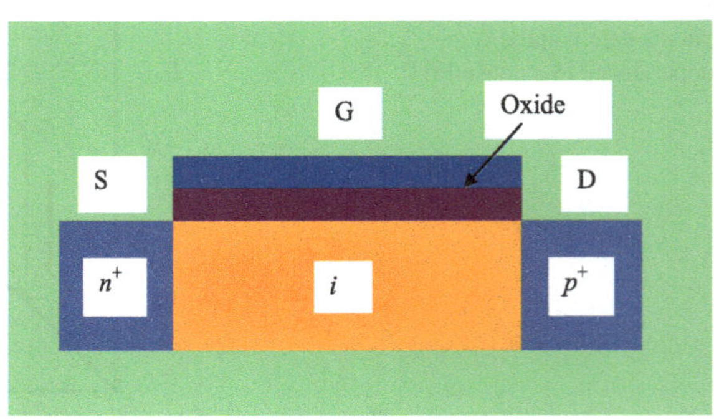

Fig. 3.56 TFET configuration

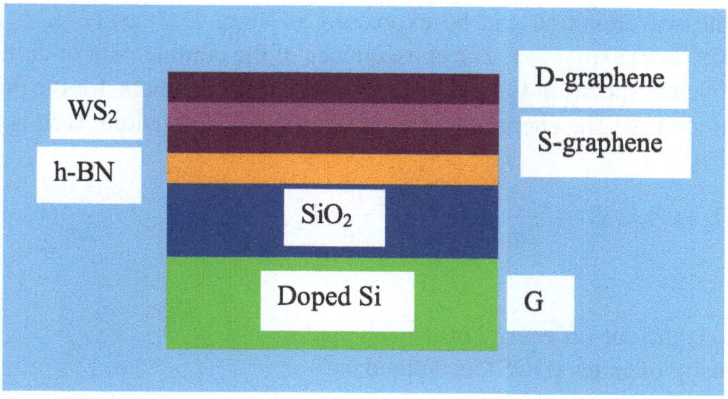

Fig. 3.57 Vertical graphene/WS_2/graphene heterostructure tunneling transistor

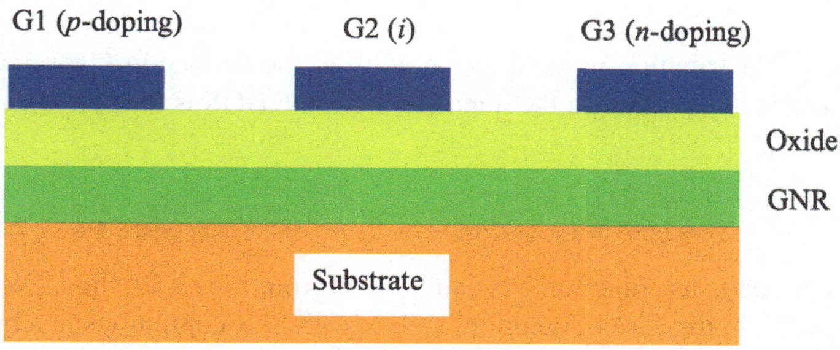

Fig. 3.58 The lateral 2D FET

3.2 Quantum Dots

Quantum semiconductor structures where the carrier motion is spatially constrained in one or more spatial directions are the building blocks of the most advanced electron devices to date. They are termed as quantum wells when the charge carriers are moving in a plane, quantum wires where the carriers are moving along a single direction (therefore, also named as quantum waveguides or electron waveguides), and quantum dots, also named as artificial atoms, when the carriers are localized on discrete energy levels.

The spatial constraints are manifested as boundary conditions associated with the Schrödinger equation. These confinements are implemented via heterostructures or directly with nanomaterials, such as single-walled carbon nanotubes, which behave as nanowires, or nanoparticles, which act as quantum dots.

In a quantum well, the carriers are confined by potential barriers in the, say, z region of width L_z, and are moving freely only in the transverse (x, y) plane, in the region with dimensions L_x and L_y.

Modeling the quantum well with infinite-height barriers, the Schrödinger equation satisfies the boundary conditions $\Psi(x, y, 0) = \Psi(x, y, L_z) = 0$ and, for

$V = 0$, the wavefunction can be expressed as $\Psi(x, y, z) = (2/L_z L_x L_y)^{1/2} \times \sin(k_z z) \exp(ik_x x) \exp(ik_y y)$. The consequence of the confinement of carrier movement is a discrete spectrum $k_z = p\pi/L_z$, with p integer numbers, for the momentum component along z. The energy dispersion relation in the quantum well is thus

$$E(k_x, k_y, k_z) = E_c + \frac{\hbar^2}{2m}\left(\frac{p\pi}{L_z}\right)^2 + \frac{\hbar^2}{2m}(k_x^2 + k_y^2) = E_{s,p} + \frac{\hbar^2}{2m}(k_x^2 + k_y^2), \tag{3.22}$$

where $E_{s,p}$ is the cut-off energy of the subband p.

The density of states (DOS), is defined as

$$\rho(E) = (2\pi)^{-3} \int_{\Sigma} \frac{dS}{|\nabla_{\mathbf{k}} E|_{E=\text{const.}}} \tag{3.23}$$

for any energy distribution in the $\boldsymbol{k}$ space, with Σ the surface in $\boldsymbol{k}$ space on which $E(\mathbf{k})$ is constant. Therefore, in the quantum well, the DOS is given by

$$\rho_{QW}(E) = \frac{m}{\pi \hbar^2 L_z} \sum\nolimits_p \vartheta(E - E_{s,p}), \tag{3.24}$$

where ϑ is the unit step function. As can be seen from Fig. 3.59, the DOS in (3.24) is discontinuous in the case of quantum wells, while it is continuous in a bulk semiconductor. When the Fermi energy is placed between the first two discrete states in the quantum well (see Fig. 3.59), the electrons inside the well form a 2D electron gas.

In a quantum wire, the carrier movement is constrained by energy potentials along two directions: y and z, but along the x direction the carriers can move without any constraint. For infinite-height potentials, the electron wavefunction

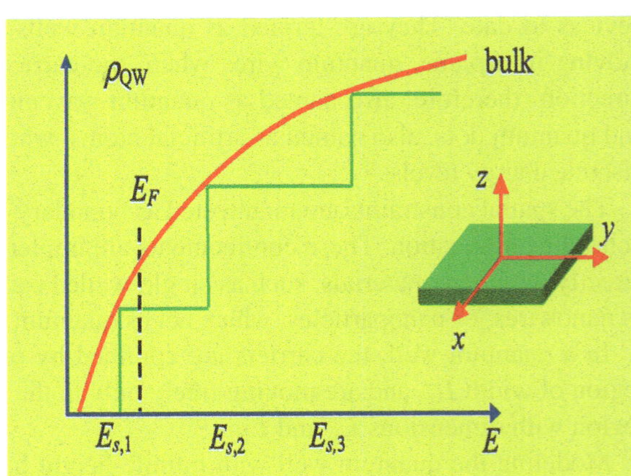

Fig. 3.59 The DOS of a quantum well

in a quantum wire with dimensions L_i, $i = x,y,z$, is given by $\Psi(x, y, z) = [2/(L_y L_z L_x)^{1/2}] \sin(k_y L_y) \sin(k_z L_z) \exp(i k_x x)$. In this case, two wavenumbers have discrete values: $k_y = p\pi/L_y$, $k_z = q\pi/L_z$, with p, q integers, the dispersion relation becomes

$$E(k_x, k_y, k_z) = E_c + \frac{\hbar^2}{2m}\left(\frac{p\pi}{L_y}\right)^2 + \frac{\hbar^2}{2m}\left(\frac{q\pi}{L_z}\right)^2 + \frac{\hbar^2 k_x^2}{2m} = E_{s,pq} + \frac{\hbar^2 k_x^2}{2m} \quad (3.25)$$

and the DOS is given by:

$$\rho_{QWR}(E) = \frac{(2m)^{1/2}}{\pi \hbar L_y L_z} \sum_{p,q} (E - E_{s,pq})^{-1/2}. \quad (3.26)$$

The DOS of the quantum wire is represented in Fig. 3.60(a).

In a quantum dot, the carrier motion is spatially confined along all directions in regions with dimensions much smaller than the mean free path and the phase relaxation length. The dispersion relation is now

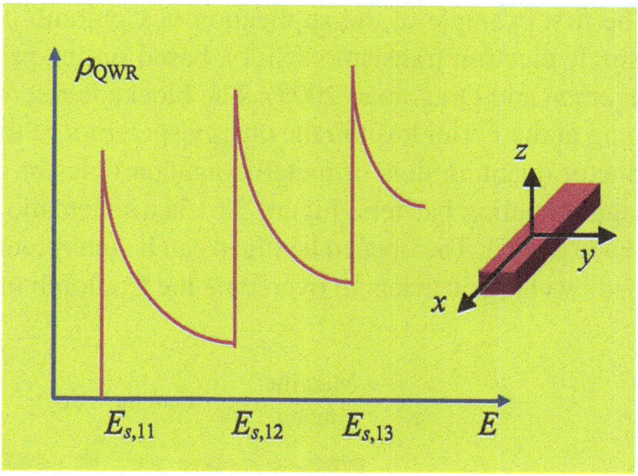

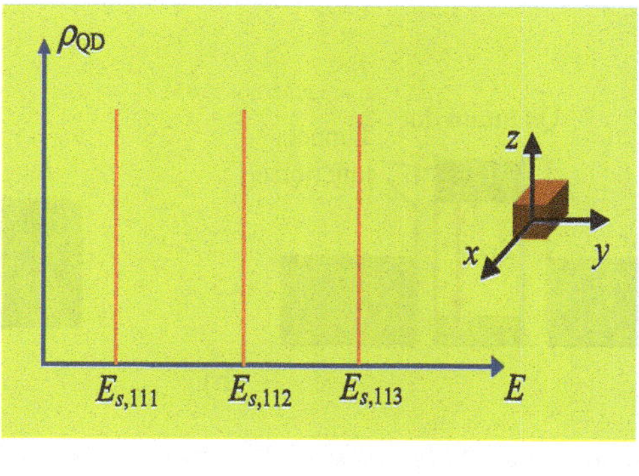

Fig. 3.60 DOS of (a) a quantum wire and (b) a quantum dot

$$E(k_x, k_y, k_z) = E_c + \frac{\hbar^2}{2m}\left(\frac{p\pi}{L_x}\right)^2 + \frac{\hbar^2}{2m}\left(\frac{q\pi}{L_y}\right)^2 + \frac{\hbar^2}{2m}\left(\frac{r\pi}{L_z}\right)^2 = E_{s,pqr}, \quad (3.27)$$

so that the DOS is proportional to the Dirac function:

$$\rho_{QD} \propto \delta(E - E_{s,pqr}). \quad (3.28)$$

and is represented in Fig. 3.60(b). Since all the energies of a quantum dot have discrete values, the quantum dot is considered as an artificial atom.

The theory of quantum dots is explained very well in the book of Harrison and Valavanis (2016), while the electronic applications of quantum dots based on semiconductors are found in the book of Wang (2012). There are also other books describing the applications of quantum dots in biology and computation (Michel 2017). Here we focus on quantum dots based mainly on 2D materials, as well as on their applications as single photon emitters, single electron transistors and in quantum computation.

In quantum dots, due to their very small geometrical dimensions, of few nanometers, the Coulomb interaction between charge carriers is strong and, in consequence, the electronic states depend on the number of charge carriers in the dot.

The first example of the application of Coulomb interaction in quantum dots is the single electron transistors (SET), based on the principle of Coulomb blockade (Dragoman and Dragoman 2009). The blockade refers to the fact that a bandgap is opening at the Fermi level in the energy spectrum of the electrons confined in semiconductor quantum dots or metallic nanoparticles or islands connected to contacts through tunneling barriers. Figure 3.61 is a schematic representation of a Coulomb blockade device. The opened bandgap can be interpreted as an extra energy, which is needed by charge carriers to overcome the Coulomb interactions, and thus to tunnel

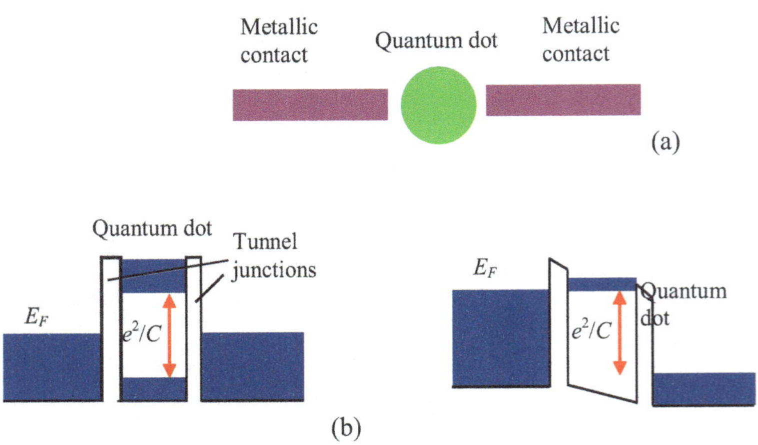

Fig. 3.61 (a) Quantum dot connected to metallic contacts via barriers. (b) The Coulomb blockade effect in the absence of an applied bias, and (c) when a DC bias of $V = e/C$ is applied

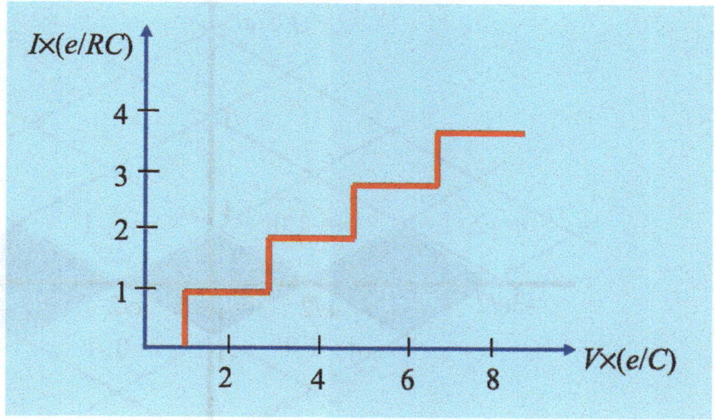

Fig. 3.62 *I–V* characteristic of the quantum dot in the Coulomb blockade regime

from one contact to the other. This extra energy is given by $e^2/2C$, with C the capacitance between the quantum dot and the contacts, and is associated to an energy gap opening of e^2/C in the electron spectrum at the Fermi level, since also the holes need the same extra energy $e^2/2C$ to tunnel into the island as the electrons (see Fig. 3.61(b)). The Coulomb blockade is observable when $e^2/C \gg k_B T$, meaning that the extra energy should satisfy the relation $e^2/C \gg \hbar/\tau$, with τ the charge carrier lifetime, and thus, $R \gg h/e^2$.

The electron tunnels into a quantum dot/island when the bandgap induced by the Coulomb blockade is overcome by the applied bias $V > e/C$. The current-voltage dependence of the quantum dot near zero bias displays a low current value. At $V = e/C$, as an electron is tunnelling in the quantum dot from one lead, the energy level in the dot is raised again by e^2/C and no further tunnelling is possible until the bias becomes $V > 3e/C$. Thus, the number of electrons on the quantum dot are raised in steps of $2e/C$, and the current-voltage dependence has a staircase-like shape (see Fig. 3.62).

The single electron transistor (SET) is similar to the structure in Fig. 3.61(a), in which the metallic contacts play the role of S and D electrodes, except for an additional gate electrode G, which enhances the charge control in the quantum dot. Such tunable quantum dots can be modeled as a parallel RC circuit where R is the tunneling resistance and C the total capacitance, equal to $C = C_S + C_D + C_G$ where the subscripts S, D, G refer to capacitances assigned to the source, drain, and gate metallic contacts of the quantum dot.

The equations which describe the tunneling into and out from the quantum dot are given by Ferry and Goodnick (1997), Dragoman and Dragoman (2009)

$$-e/2 \mp [en + (C_G + C_D)V - C_G V_G'] > 0, \tag{3.29a}$$

$$-e/2 \pm [en - C_S V - C_G V_G'] > 0, \tag{3.29b}$$

where $V_G' = V_G + Q_p/C_G$, with Q_p a parasitic charge.

The above inequalities determine a family of straight lines in the (V, V_G) plane, intersecting each other, as can be seen in Fig. 3.63. These curves consist of rhomboidal

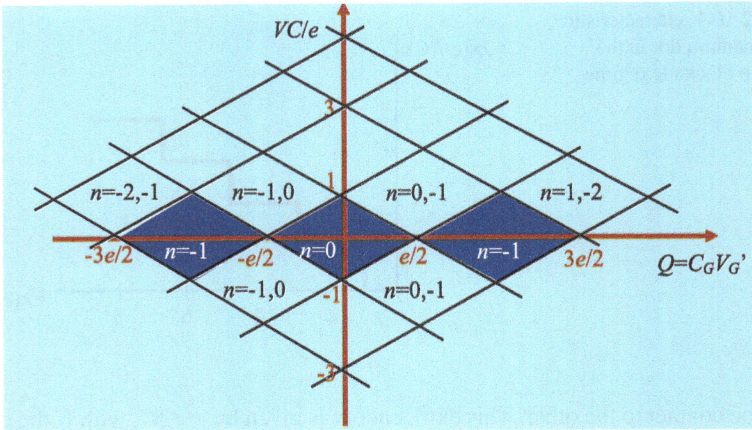

Fig. 3.63 The Coulomb diamonds

Coulomb blockade regions, termed Coulomb diamond regions, and inside these regions there is a finite, fixed number of carriers. Outside the Coulomb diamonds, the number of carriers is no longer fixed, but varies with integer numbers. The dependence of the drain voltage versus gate voltage is represented in Fig. 3.63

A first example of a 2D material quantum dot consisting (Fig. 3.64) of the carving of a graphene sheet is shown in Fig. 3.65 (Barreiro et al. 2012). The carved graphene structure is transferred on a SiO_2 substrate grown on a doped Si layer that plays the role of a gate. The dimensions of the constriction are in the range 1–2 nm. There is no lithographic technique able to obtain such small features, so that the graphene quantum dot is fabricated by electro-burning in air. A ramp voltage (1 V/s) is applied to the graphene ribbon and the current is monitored at every 200 μs, the conductance is measured, and the process is reloaded when the conductance is decreasing 10% of its value during 200 mV of the voltage ramp. The current is decreasing progressively until low values are attained and the quantum dot is formed (see Fig. 3.66).

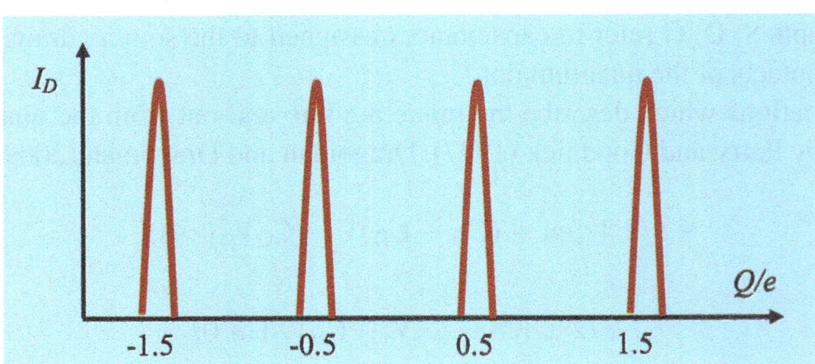

Fig. 3.64 The current through a quantum dot connected via tunneling contacts as a function of the charge/gate voltage

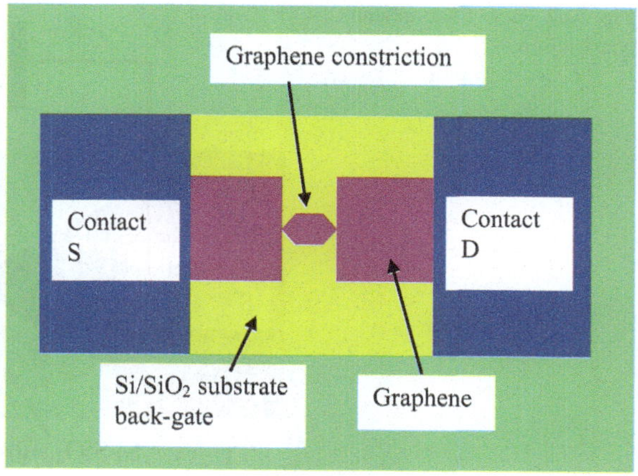

Fig. 3.65 Carved quantum dot in graphene

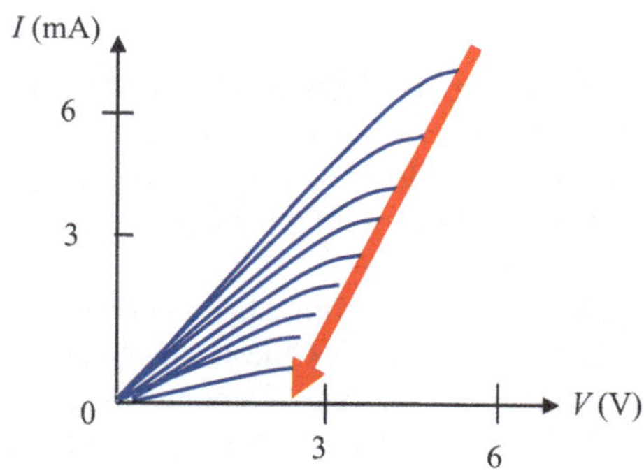

Fig. 3.66 *I–V* shaping for a quantum dot formation in graphene (after Barreiro et al. 2012)

In the case of 2D materials, quantum dots and Coulomb diamonds are retrieved, generally, at low temperatures. In (Barreiro et al. 2012) it was reported that in carved graphene quantum dots described at the beginning of this section, a Coulomb diamond was observed for back-gate V_G ranging between –50 and +50 V and a voltage V ranging between –1 and +1 V at 10 K. A quantized level $\Delta E = 0.8\,\mathrm{eV}$ was extracted from these measurements. The Coulomb diamond height on the V axis is measuring the energy needed to add a carrier to the quantum dot, energy equal in this case to 1.6 eV, which is a rather high value.

In graphene nanoconstrictions as those above, interference effects appear in the conductance (Gehring et al. 2016). In this respect, a bow-tie graphene channel was patterned between two metallic contacts, as illustrated in Fig. 3.67. The substrate is SiO_2 grown over a *p*-type doped Si working as a back gate. The length of the bow-tie is 4 μm, while its minimum width (neck) is 200 nm. Then, an electroburning process is applied, as in the example above. The neck of the bow-tie graphene is narrowed, while the resistance of the nanoconstriction is increasing, so that a threshold resistance should be imposed, of 500 MΩ in this case.

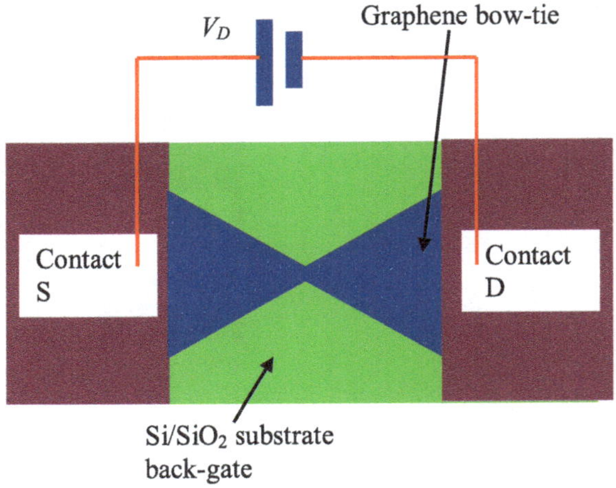

Fig. 3.67 Bow-tie graphene constriction

The transport regime is able to show when a gap is formed or when a quantum dot is formed. When a gap forms, the current is assigned to the tunneling process, and there is no dependence of V_D on V_G. The dependence of the current on V_D is that of a tunneling current, typical for a rectifier (see Fig. 3.68). The gap has the size of few nanometers.

When no gap is formed, a quantum dot appears and its signature in the current-voltage dependence is the disappearance of rectification (see Fig. 3.69) and the occurrence of Coulomb diamonds in the V_D–V_G dependence.

The conductance of strongly coupled quantum dots has as signature a chess-board-like form due to interference patterns, meaning that the carriers are transported ballistically between the quantum dots. The energy spacing is about 4–5 meV at 4 K and the coherence length (L) of carriers is greater than 800 nm, so that resonances occur at $k_F L = n\pi$. The conductance dependence on V_G shows minima and maxima, as displayed in Fig. 3.70.

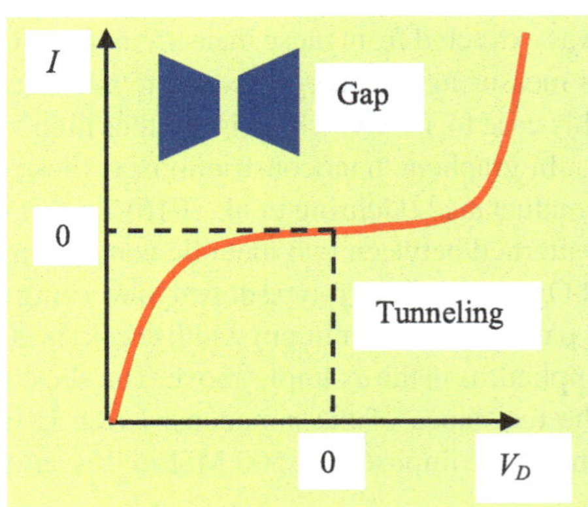

Fig. 3.68 Current-voltage signature of the occurrence of the graphene nanogap

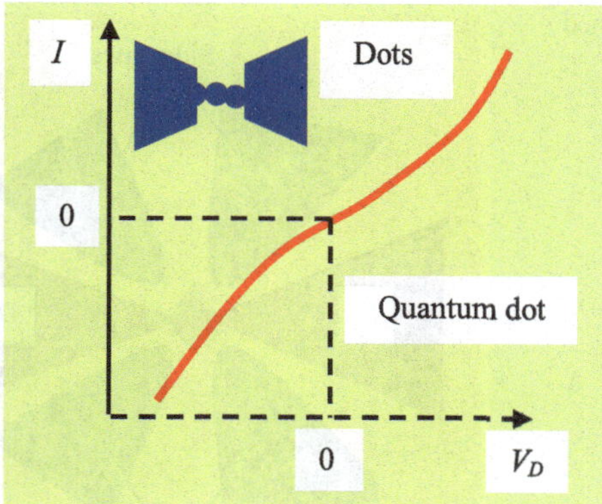

Fig. 3.69 Current–voltage dependence signature of the quantum dot regime

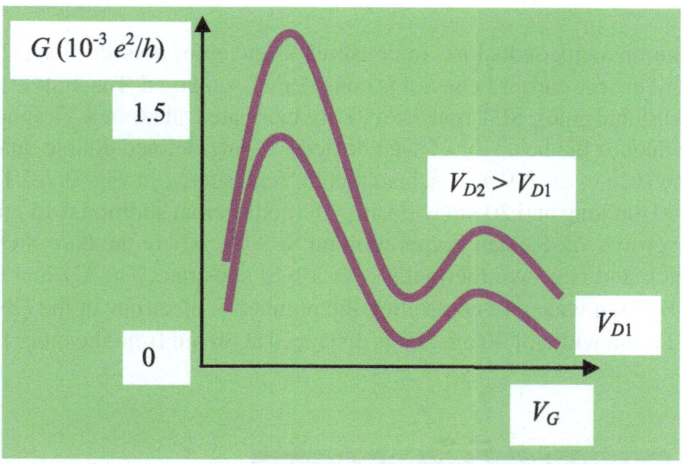

Fig. 3.70 Interference signature in quantum dots

Quantum dots can be fabricated via e-beam techniques and using side gates to confine the carriers in all dimensions, since the carriers in graphene form a native 2D gas. There are two main reviews about etched quantum dots via e-beam techniques, where the physical aspects of graphene quantum dots are explained in detail (Güttinger et al. 2012; Bischoff et al. 2015). A quantum dot with side gates is shown in Fig. 3.71.

Discrete energy levels and Coulomb blockade effects are evidenced in such structures at low temperatures, depending on the voltage applied on different side gates (Bischoff et al. 2015). The role of side gates was explained in the previous chapter. In fact, the behavior of the above device is modeled as three quantum dots in series: two outer dots in the two constrictions and an inner quantum dot, in the island. The strength of the coupling between the dots can be tuned via the side-gate potentials;

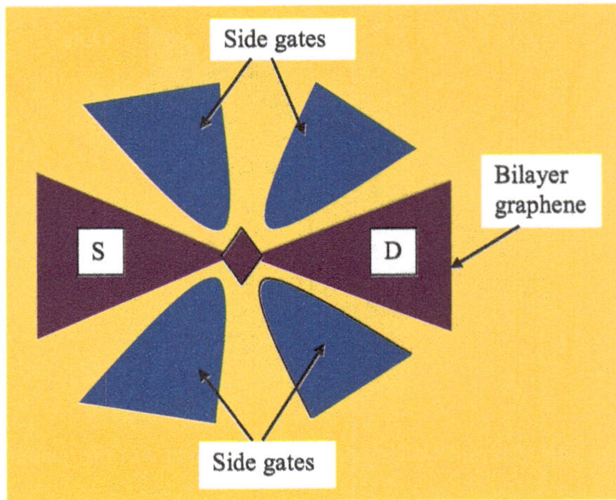

Fig. 3.71 Side-gate-defined quantum dots in bilayer graphene

unlike common semiconductors, in constrained quantum dots in bilayer graphene the wealth of charge carrier behavior is considerably enlarged (Bischoff et al. 2015). Adding additional gates, SET transistors were fabricated and tested. Graphene edges strongly influence the behavior of such devices. A gate-defined double quantum dot in graphene (Liu et al. 2010) is schematically represented in Fig. 3.72. The GNR, which is 800 nm long and 20 nm wide, is patterned over an additional 15 nm of thick SiO_2 layer grown via e-beam evaporation on Si/SiO_2, where the bare SiO_2 layer is 285 nm thick and is grown thermally over a *p*-Si substrate. The G1 and G3 gates, which are 600 nm wide, are controlling the number of electrons in the GNR, while the gate G2 has a width of 40 nm and is distanced at 80 nm from the other two gates,

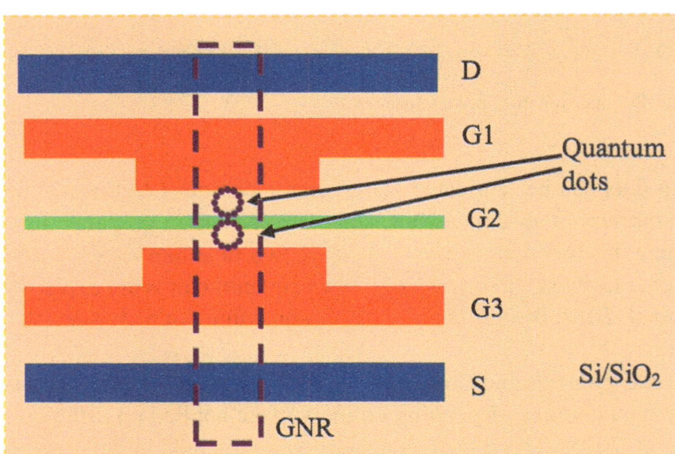

Fig. 3.72 Gate-defined double quantum dots

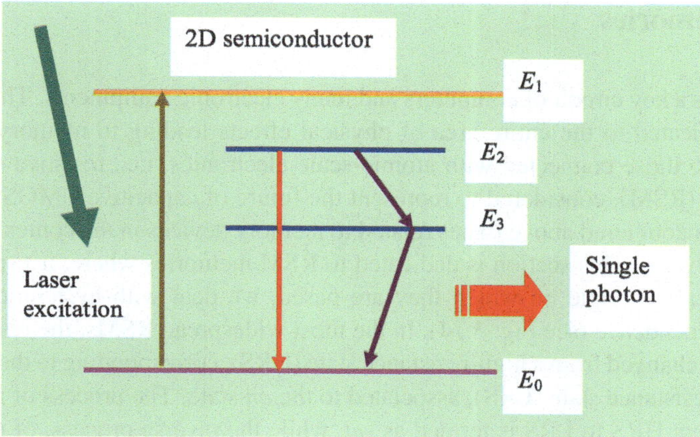

Fig. 3.73 Single-photon generation from color centers

its role being that of controlling the coupling between quantum dots. Measurements are made at 50 mK, and the features of the two dots are evidenced.

In many 2D materials, deep in-gap states (color centers) exist, which are highly localized states, analogous to atoms or quantum dots. This is the case of h-BN, with a bandgap of 6 eV, where excitation with a laser at $\lambda = 535$ cm has produced the emission of single-photons at room temperature—a crucial step forward toward quantum computing based on photons (Tran et al. 2016). The h-BN is transferred on Si and the experiments are performed at room temperature. The single-photon light is transmitted then into an optical fiber and further used in a photonic circuit (a Hanbury-Brown and Twiss interferometer) to test its properties. Figure 3.73 illustrates the transitions taking place in h-BN, assuming a three-level model, the photons being emitted by transitions to the ground state E_0 from the metastable state E_3, which is associated to color centers; E_1 denotes the excited state and E_2 another discrete level in the bandgap of h-BN.

There are other materials able to produce single photons at room temperature due to color centers, such as diamond, SiC, arsenide quantum dots, or carbon nanotubes (for a review see Aharonovich et al. 2016). Quantum dots were obtained in yet other 2D materials-like structures, in the form of nanoparticles embedded in various chemical solutions. Such quantum dots are useful in biology and energy storage, but they cannot be used in nanoelectronics because of their poor controllability and reproducibility. Even the above described quantum dots and devices are not highly reproducible due to edge effects.

3.3 Memories

Memory is a key circuit of computers and many electronic equipments. This section is not dedicated to the entire area of physical effects leading to memory devices, but only to those connected with atomic-scale electronics, i.e., resistive-switching memories (RSM), considered to represent the future of capacitive CMOS memory. We have encountered above issues related to memory devices in the context of ferroelectric devices. This section is dedicated to RSM memories where, irrespective of the physical principle on which they are based, we deal with hysteretic current-voltage dependence (see Fig. 3.74). In the most widespread RSMs, the resistance is reversibly changed from a high-resistance state (HRS), corresponding to the *off* state, to a low-resistance state (LRS), associated to the *on* state. The process of resistance change from HRS to LRS is termed as *set*, while the inverse process, of resistance change from LRS to HRS is termed *reset*.

There are two large classes of RSMs, in which the switching can be either bipolar or unipolar. This classification is based on the fact that the *set* and *reset* processes can be performed using the same polarity of the voltage (unipolar) or with two different voltage polarities (bipolar). In Fig. 3.74, we are dealing with a unipolar RSM, while Fig. 3.75 illustrates a bipolar RSM. There are also different physical systems leading to RSM. We analyze here only a couple of them; a recent review on RSMs can be found in Slesazeck and Mikolajick (2019).

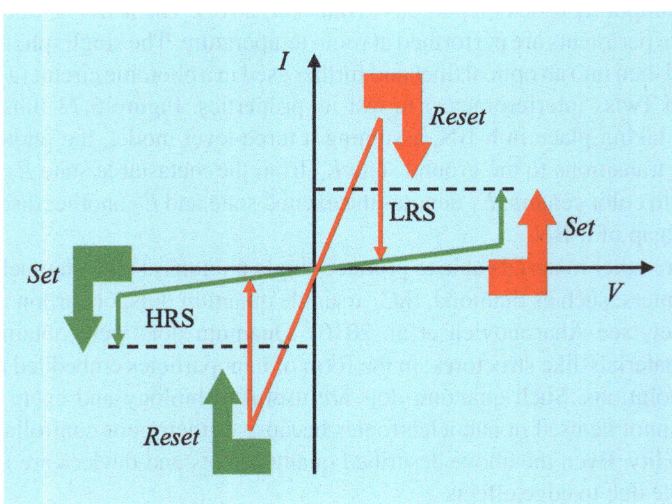

Fig. 3.74 Current–voltage dependence of the RSM (unipolar)

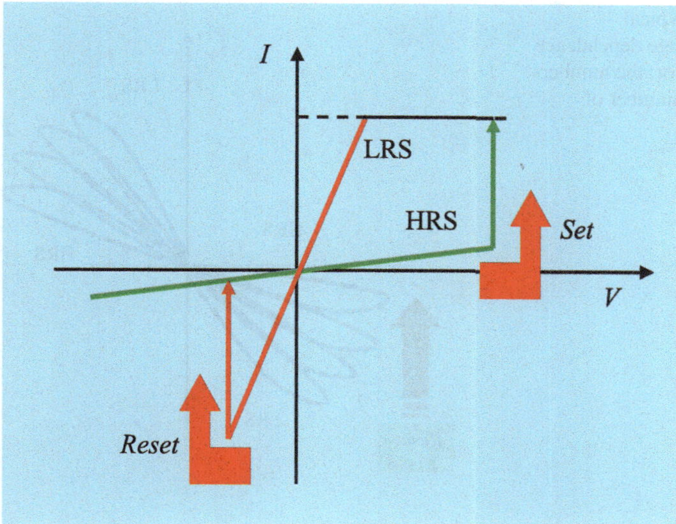

Fig. 3.75 Current–voltage dependence of the RSM (bipolar)

3.4 Memristors

3.4.1 Filament Memristors

The memristor is a resistance with memory. As such, the memristor can be considered as the fourth fundamental circuit element, playing a similar role as the inductance (L) and capacitance (C) in the circuit theory. The experimental evidence of the existence of memristors has arrived decades the concept was put forward, when technology has reached its maturity at the nanoscale.

The memristor is always a nonlinear circuit element, showing a current-voltage characteristic with a unique imprint: a pinched hysteretic behavior (Chua 2014a) (see Fig. 3.76). If more sweeps are performed, the LRS is progressively increasing while the HRS is decreasing.

From Fig. 3.76 it follows that the resistance of the memristor depends on the history of the applied voltage, memorizing thus its previous state when the excitation is off. As such, the memristor is a non-volatile memory, mimicking the behavior of a brain synapse (Jo et al. 2010).

The first experimental evidence of the memristor was made using a TiO_2 MIM structure, which was doped over a distance *w* (Strukov et al. 2008). The physical explanations of the behavior of these memristors were given in Chap. 2. Since then, a lot of work was invested in searching for memristive behavior in other oxides, crossbar arrays were fabricated and learning mechanisms were implemented. Moreover, new physical mechanisms for memristors were discovered.

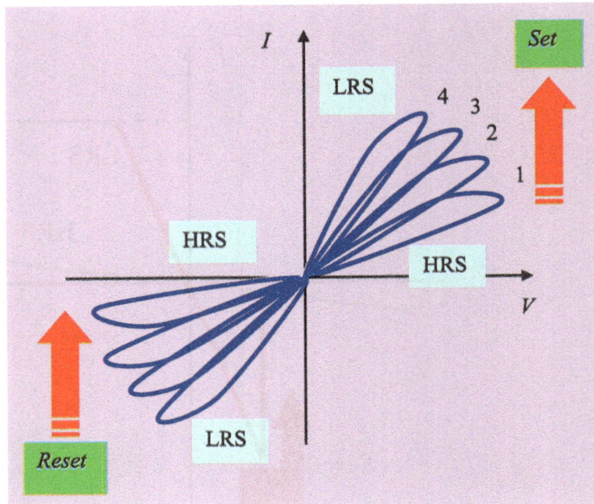

Fig. 3.76 Typical current–voltage dependence of a memristor; the numbers indicate the number of sweeps

The most encountered category of memristors is valence change memristors (VCM) or filament memristors, which are MIM-like structures, consisting of top and bottom electrodes from the same metal or from different metals, and a very thin binary oxide sandwiched between them, as shown in Fig. 3.77. Among the many oxides used for memristors, the most known binary oxides are TiO_x and HfO_x. There are many reviews (see, for example, Hu et al. 2014) and books (see for instance Tetzlaff 2014) regarding VCMs.

How is the filament in a memristor formed and how is it evolving in the *set* and *reset* processes? These questions can be answered by noting that, in the *set* process, oxygen vacancies pairs (anyons) are created inside the oxide due to the impact ionization process:

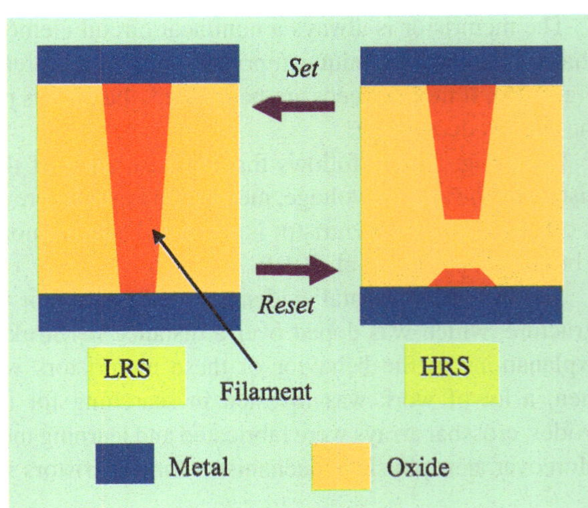

Fig. 3.77 The filament memristor and its states

$$O_L \rightarrow V_O^{2+} + O_i^{2-} \tag{3.30}$$

where O_L is the oxygen in the oxide lattice, V_O^{2+} the oxygen vacancy and O_i^{2-} stand for oxygen ions created additionally in the lattice. The O_i^{2-} ions are migrating and agglomerating at the anode (top electrode), forming bubbles which can be seen via TEM methods (Kwon et al. 2010) and which could damage the anode according to the reaction

$$O_i^{2-} \rightarrow 2e^- + O_2 \uparrow \tag{3.31}$$

If the O_i^{2-} ions migrate toward the anode, the oxygen vacancies are migrating in the opposite direction, at the cathode (bottom electrode) and trap the electrons:

$$C^{z+} + ne^- \rightarrow C^{z-n} \tag{3.32}$$

where C^{z+} are cations. Thus, at the cathode the population of oxygen vacancies is formed and propagates to the anode, developing current filaments, which generate conduction in the oxide and hence the transition from HRS to LHS.

In the *reset* process, on the other hand, large densities of currents in the oxide with a thickness of 6–15 nm are cancelling oxygen vacancies via Joule heating, according to the reaction:

$$O_i^{2-} + V_O^{2+} \rightarrow O_L \tag{3.33}$$

where O_i^{2-} are oxygen ions. So, the LHR is transformed into HRS.

According to Slesazeck and Mikolajick (2019), there are several problems related to VCMs:

- the LRS (*on* state) can be controlled by current compliance, limiting the value of the current,
- the write current is high and forms a filament, which is stable and able to drive the VCM in a LRS state.

All RMSs arranged in crossbar arrays need selectors, which are transistors, diodes or other nonlinear devices that enable the desired RMS cells and disable other cells.

3.4.2 Semiconductor Memristors

A new development in memristors implies using nanosized/atomically thin semiconductors, avoiding thus electrochemical reactions and their unavoidable uncertainties. Several examples are presented below, starting with GaN nanomembranes. For instance, GaN memristive devices were obtained using a modified version of the surface charge lithography (SCL), which involves ion-beam-writing of surface negative charge followed by photoelectrochemical (PEC) etching of the GaN layer (Dragoman et al. 2016b). An ultrathin GaN membrane array with dimensions of 12 $\times$ 184 μm^2 was produced in this way, by treating some regions with 0.5 keV Ar^+ ions at the dose of 10^{11} cm^{-2}. In the ion-treated regions conductive GaN membranes are formed, as displayed in the SEM image in Fig. 3.78. These GaN ultrathin membranes, with thickness of 15.6 nm, are suspended over a network of nanowires representing threading dislocations.

The current-voltage dependence of this structure was measured by sweeping repeatedly the voltage from −5 V up to +5 V and then backwards. Each sweep is

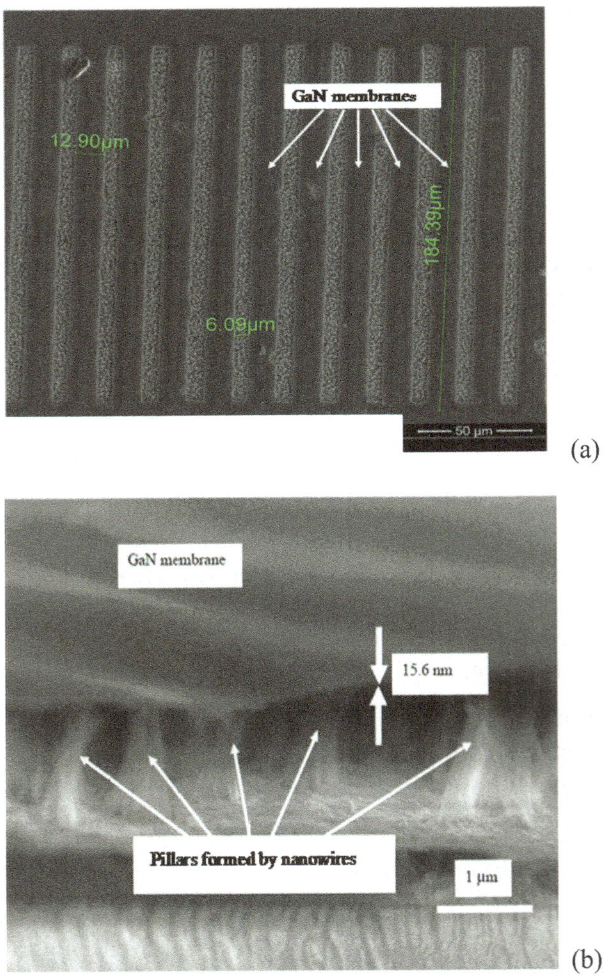

Fig. 3.78 SEM images of (a) the GaN membrane array and (b) its cross-section (Dragoman et al. 2016b)

denoted by 1, 2, 3, 4, in Fig. 3.79. This figure shows that the GaN utrathin membrane is acting indeed as a memristor, having a pinched hysteretic shape.

To identify the physical effects which produce the memristive behavior, we have represented in logarithmic scale the current-voltage dependence, and fitted it with the relation $I = V^{\alpha}$ (see Fig. 3.80). The first observation is that the α values are the same for the first and the fourth sweep, indicating that the conduction mechanisms are the same.

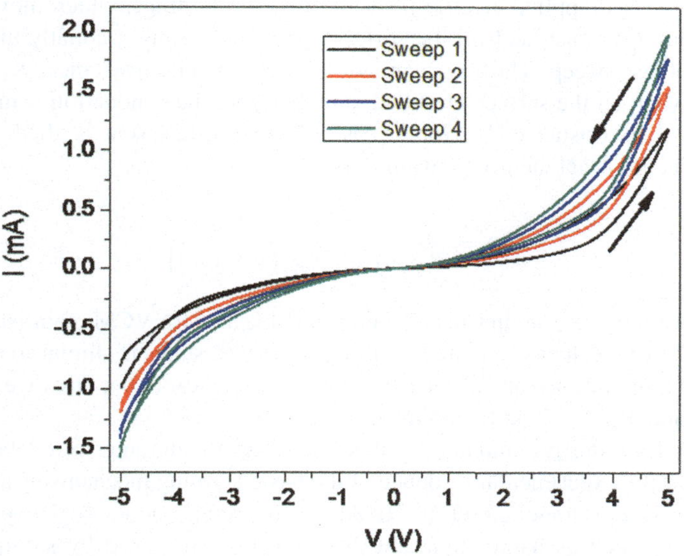

Fig. 3.79 The GaN ultrathin membrane memristor (Dragoman et al. 2016b)

Fig. 3.80 Log(I) versus log(V) dependence of the first and fourth sweep (Dragoman et al. 2016b)

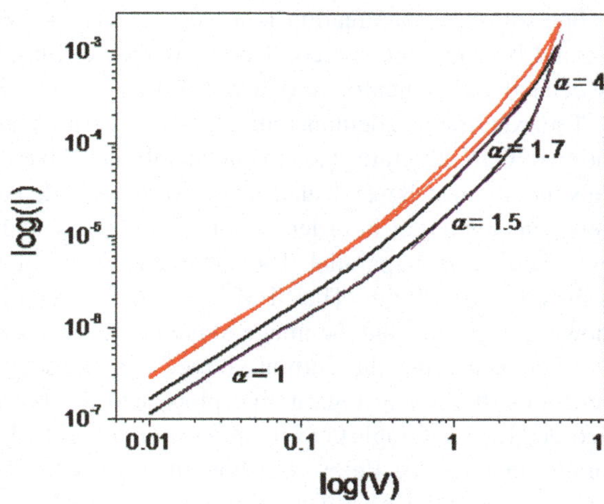

However, the α value is not constant in a sweep. Different values of this parameter suggest different conduction mechanisms occurring in the device. More precisely, for low-voltages $\alpha = 1$, meaning an Ohmic behavior, i.e., a linear dependence between current and voltage. At higher values of the voltage, on the other hand, the I–V dependence is becoming nonlinear and α increases up to 4. The space charge limited current mechanism was identified as responsible for such a large value of α. This mechanism takes place since the negative-charged deep traps produced during irradiation with ions in the fabrication process of membranes boost the applied field.

This increased applied electric field is acting on the traps, which move toward the membrane surface, and fill the surface states, increasing gradually the current at each voltage sweep. Thus, we can associate an *on*-state resistance, R_{on}, for the situation when all the surface are filled, and therefore the conduction is maximum, and an *off*-state resistance R_{off} for the case when no surface state is filled. Then, we can describe the GaN memristor using formula

$$i(t) = \frac{v(t)}{R_{on}\gamma(t) + R_{off}[1 - \gamma(t)]} \tag{3.34}$$

This formula was used in Chap. 2, being suitable also for VCM memristors. Note that, the above mechanism related to filling of surface states is similar to the VCM mechanism, of presence and absence of oxygen vacancies. In our case, we have $R_{on} = 806\ \Omega$ and $R_{off} = 77\ \text{k}\Omega$, with the ratio $R_{off}/R_{on} = 95$.

Starting from this preliminary result, several GaN membranes were fabricated in parallel, for the experimental evidence of the basic learning mechanisms associated with memristors (Dragoman et al. 2018d). In particular, non-associative learning mechanisms, such as habituation and dishabituation, followed by storage of the response to a certain electrical stimulus in array of parallel GaN memristors were demonstrated, in analogy with living beings.

Habituation is a learning mechanism, dedicated to disregarding irrelevant stimuli. In opposition, dishabituation is a learning process able to respond to a stimulus ignored by habituation. Leon Chua has theoretically predicted that habituation is associated with a memristor (Chua 2014b).

The experimental demonstration of these learning mechanisms involved the fabrication of GaN structures having one membrane, 2 parallel membranes and 3 parallel GaN membranes. Several such arrays were fabricated and then the metallic contacts were selectively cut in order to obtained devices with one memristor, and two or three memristors in parallel. The metallic contacts are Ti/Au having the thicknesses of 50 nm and 200 nm. The SEM image of a three-parallel memristor structure is shown in Fig. 3.81 and the image of the entire chip is displayed in Fig. 3.82.

After measuring the current-voltage dependence, the conduction mechanisms were identified using logarithmic plots, and the behavior of current in time was recorded for: (i) a single GaN memristor, (ii) two-parallel memristors, and (iii) three-parallel memristors. Below we show the results for three-parallel memristors only (see Fig. 3.83); the other data can be found in Dragoman et al. (2018d).

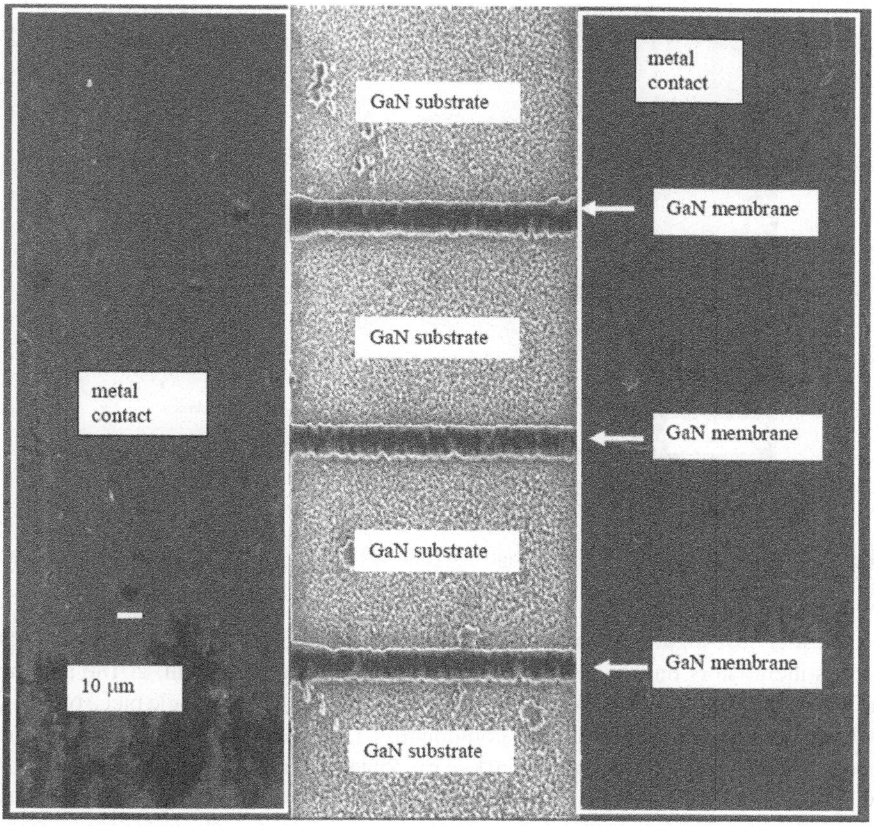

Fig. 3.81 SEM image of a three-parallel GaN membrane memristors (Dragoman et al. 2018d)

The habituation-dishabituation learning mechanisms can be observed from the current dependence in time. From Fig. 3.83(c) it can be seen that at the voltage excitation of +6 V the current reaches a steady state, but increasing the number of sweeps in time, the current between two consecutive sweeps is decreasing up to a certain sweep after which the currents for the following sweeps are superimposed and memorized—the habituation is accomplished. Only when another bias voltage is able to change the state of the circuit, in this case −6 V, dishabituation is taking place. In this situation, the signal changes its shape, but by sweeping in time, again, we are reaching another habituation state. The habituation learning mechanism has as physical explanation the gradual filling of surface traps due to the movement of trapped negative charges to the surface states, whereas dishabituation is the result of the change of the electrostatic profile inside the GaN membranes. This behavior was found in all memristor configurations, but when the number of memristors in parallel is increasing, the learning time is reduced by 30% in comparison to a single memristor at the same voltage excitation.

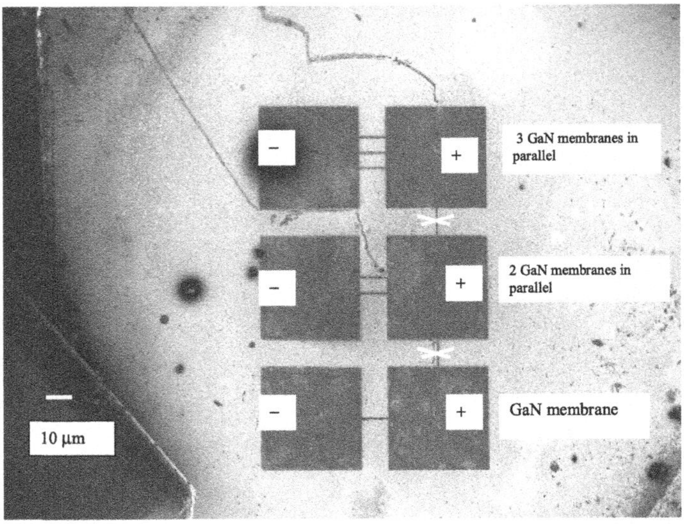

Fig. 3.82 The GaN memristor chip (Dragoman et al. 2018d)

Further investigations concerned photomemristors, i.e., nonlinear devices that act as memristors only if illuminated by light (see Dragoman et al. 2018e and the reference therein). Such a device can be implemented with just a single piece of SnS_2, which is a layered semiconductor compound—a van der Waals material, with the area of 1 cm^2 and the thickness of 100 μm, contacted laterally by silver electrodes with the area of 1 mm^2. The SEM image in Fig. 3.84 shows that SnS_2, which acts naturally as a memristor device when illuminated by a sun simulator, is formed from stacks of monolayers, each of them having the thickness of about 0.6 nm.

When the SnS_2 sample is illuminated with 1.1 sun, the current-voltage dependence has the form illustrated in Fig. 3.85, which shows a typical memristive behavior at illumination. The linearity between photocurrent and voltage (Fig. 3.85) and between photocurrent and light power (Fig. 3.86) indicates that the mechanism responsible for the memristive behavior is a slow collection of photogenerated carriers.

Memristors based on 2D semiconductors have been explored intensely in the last years; for an updated review see Zhang et al. (2019). The large majority of 2D materials memristors are based on MoS_2, although there are memristors based on h-BN, $MoTe_2$, WSe_2, or WS_2. The interest lies mainly in solid-state memristors that can be replicated at the wafer level, and the only 2D materials which are grown today even at the 300 mm wafer scale are MoS_2 and WS_2.

There are two types of memristors based on 2D materials: two-terminal memristors and three-terminal memristors, termed also as memtransitors. The 2D two-terminal memristors are MSM vertical structures, formed from an atomic-thin semiconductor sandwiched between two metals. In this case, the memristive effect is due to charge trapping or space-charge effects. An example among many is the 2D heterostructure graphene/$MoS_{2-x}O$/graphene (Wang et al. 2018). This memristor is entirely fabricated from 2D atomic thin materials, showing an *on/off* ratio of 10^7 and having maximum currents beyond 1 mA level. In addition, it works even at 340 °C (see Fig. 3.87).

Nonvolatile memories based only on 2D materials can be implemented also. An example in this sense is the heterostructure MoS_2/graphene, having resistive electrodes (Bertolazzi et al. 2013). This memory shows four orders of magnitude difference in current between *program* and *erase* states, and is designed to have a floating gate, which is isolated by dielectrics from MoS_2 and the metallic electrodes (see Fig. 3.88). The drain current versus gate voltage dependence shows hysteretic

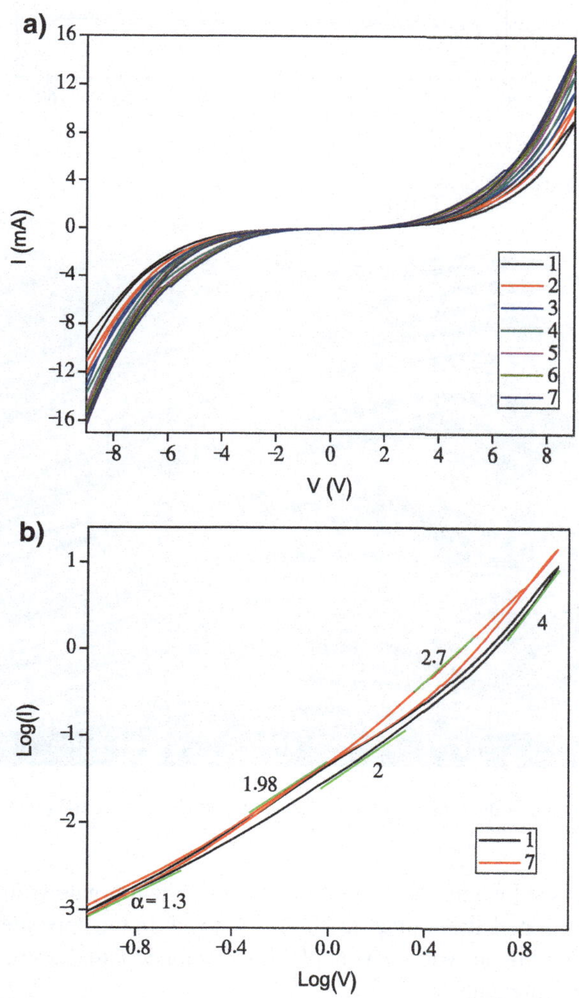

Fig. 3.83 Current-voltage dependence, at different sweeps, of three memristors in parallel: (a) linear, (b) logarithmic scales. (c) Current-time dependence for a voltage of +6 V (solid lines) and −6 V (dotted lines) (Dragoman et al. 2018d)

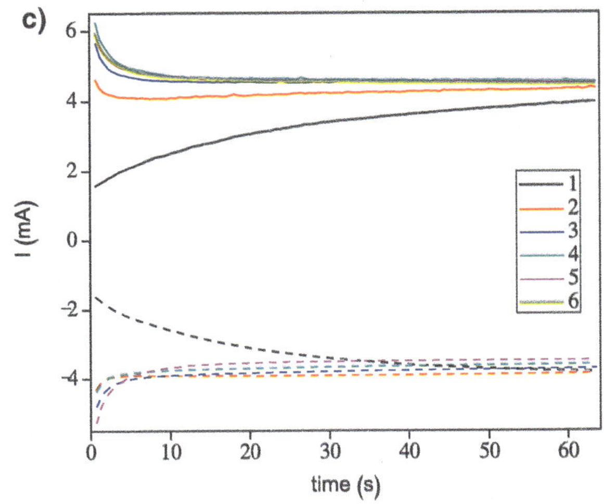

Fig. 3.83 (continued)

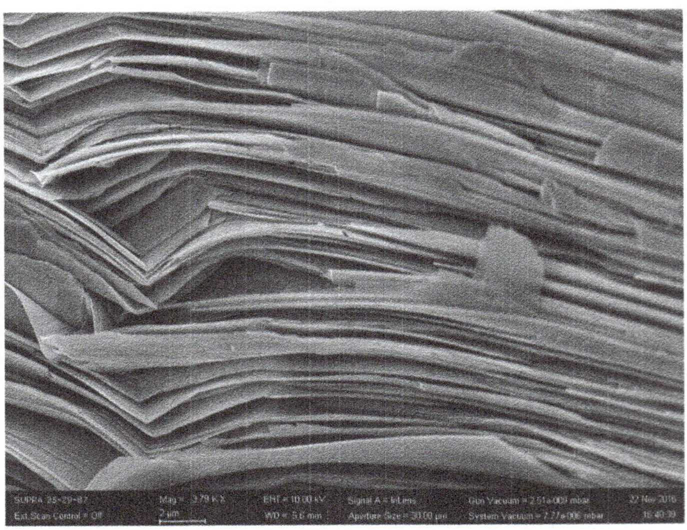

Fig. 3.84 SEM image of the SnS_2 layered materials (Dragoman et al. 2018e)

behavior, expressed by a memory window of 8 V. If the top-gate voltage is positive, the carriers are trapped after tunneling the MoS_2 layer. In the opposite situation, i.e., when a negative top-gate voltage is applied, the trapped carriers are released from the graphene floating gate.

In the last years, memtransistors were developed based on MoS_2 grown by CVD, which has a polycrystalline structure, and memristive effects are due to the modulation of the Schottky barrier at the source and drain electodes of the MoS_2 back-gate

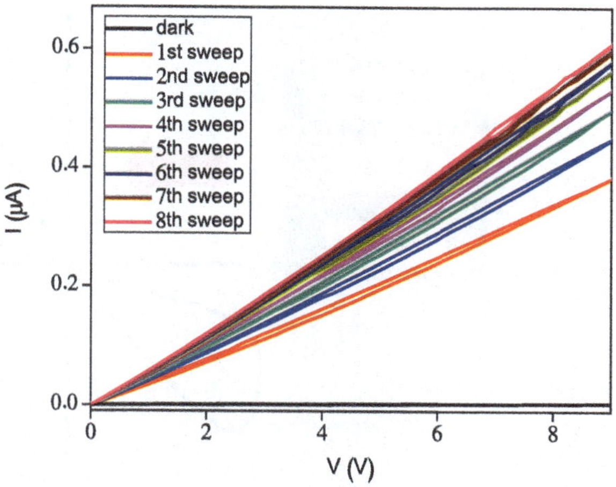

Fig. 3.85 The SnS_2 memristor (positive voltage bias) (Dragoman et al. 2018e)

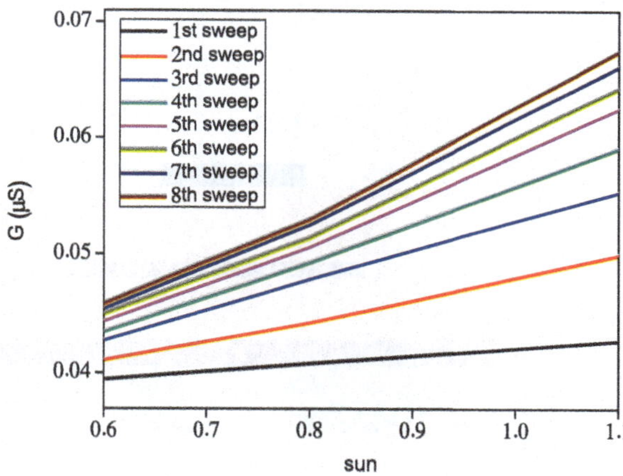

Fig. 3.86 Conductance of the photomemristor versus light intensity (Dragoman et al. 2018e)

transistor, caused by the migration of sulfur vacancy defects (Sangwan et al. 2018). The surface of polycrystalline MoS_2, with a thickness of 2–9 layers depending on the growth conditions, shows grain boundaries with dimensions up to 3 μm, and these grain boundaries are intermediating the migration of sulfur vacancy defects towards the FET terminals (see Fig. 3.89) when a drain-source voltage is applied. The memtransistor is represented in Fig. 3.90.

Graphene boundaries of the order of 1 μm and even bigger are also typical for graphene grown by CVD at the wafer level. Irrespective of the material, inside the grains a monolayer or few-atomic layers of MoS_2 or graphene can be found, while at grain boundaries there are multilayers with various thicknesses. The defects are present anywhere on the surface of the 2D materials grown by CVD, but mainly on the boundaries.

The memtransitors have several advantages over 2D two-terminal memristors: the electroforming process, used in filament memristors, is not necessary, and the

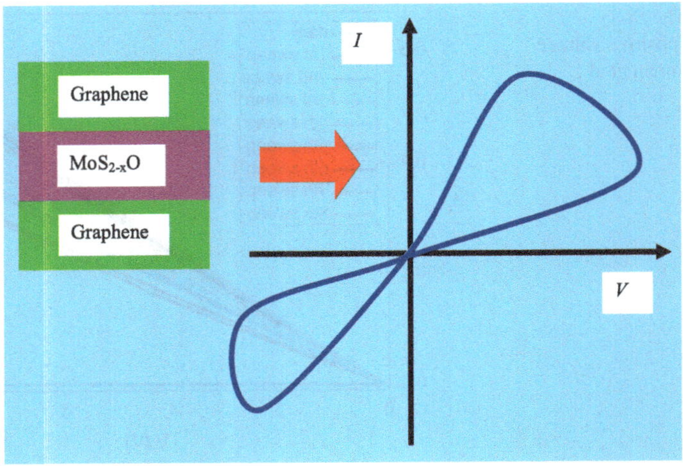

Fig. 3.87 Memristor formed only from 2D atomic layered materials

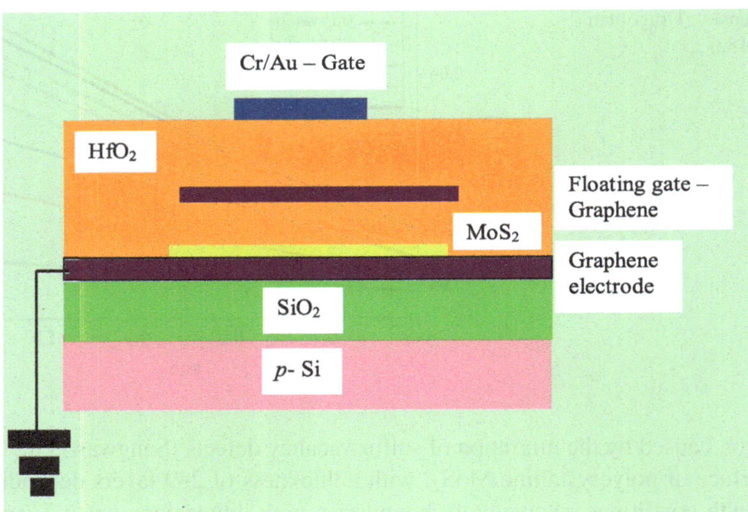

Fig. 3.88 Nonvolatile memory based on MoS_2/graphene

devices have tunable memory functions and high *on*/*off* ratios. However, to benefit from these advantages, initially the drain voltage applied on memtransistors was 80 V (Sangwan et al. 2018), which is a too high value to be accepted in nanoelectronics today. Various solutions were proposed to circumvent this problem, such as the use of a top gate (Wang et al. 2019a) or localized beam irradiation (Jadwiszczak et al. 2019). However, the drain voltage has not decreased under 12 V until now. This is still a large drain voltage for memories.

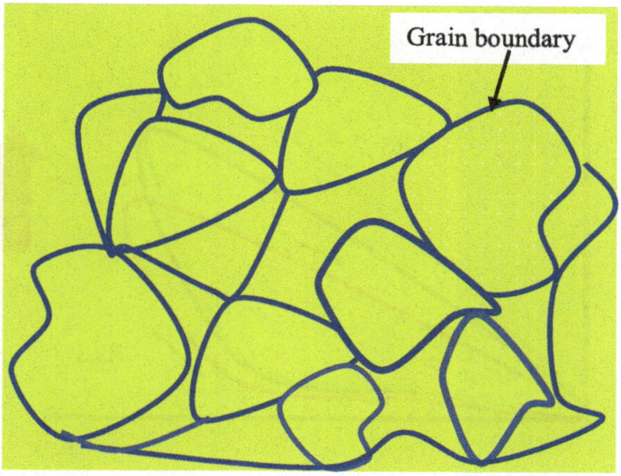

Fig. 3.89 Schematic representation of grain boundaries in polycrystalline MoS_2

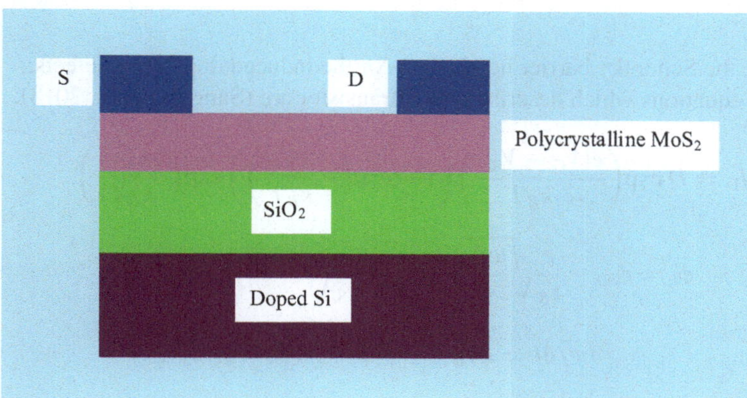

Fig. 3.90 The polycrystalline MoS_2 memtransitor

The typical memtransitor drain current-drain voltage dependence is shown in Fig. 3.91. This behavior can be described using the state-variable equations of memristors (Sangwan et al. 2018):

$$\begin{aligned} \mathrm{d}w/\mathrm{d}t &= f(w, V, t) \\ I_\mathrm{D} &= g(w, V_\mathrm{G}, V_\mathrm{D}) \end{aligned} \tag{3.35}$$

where w is the state variable of the memtransistor, related to the doping of the source and drain Schottky contacts by

$$\phi_b \approx \sqrt{w \Delta n} \tag{3.36}$$

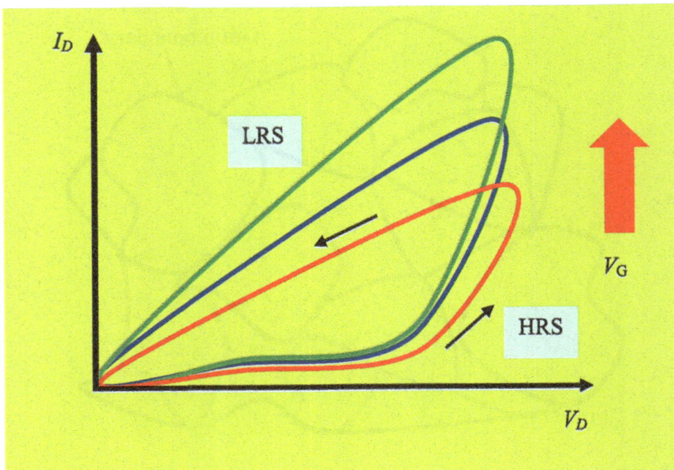

Fig. 3.91 Typical $I_D - V_D$ dependence of memtransistors

with ϕ_b the Schootky barrier height and Δn the induced doping by defects.

The equations which describe a memtransistor are (Sangwan et al. 2018):

$$I_D = D\exp\left(\frac{e(V_G - V_{th})}{c_1 k_B T}\right)\left(1 - \exp\left(-\frac{eV_D}{k_B T}\right)\right)\exp\left(\frac{\phi_b(w)}{k_B T}\right) \tag{3.36}$$

$$\phi_b = \phi_{b0} - \frac{e}{\varepsilon_S}\sqrt{\frac{w\Delta n}{4\pi}} + \sqrt{\frac{e}{4\pi\varepsilon_S}}\left(\frac{2en(\phi_{b0} + A|V_D|)}{\varepsilon_S}\right)^{1/4} \tag{3.37}$$

$$\partial w/\partial t = EI_D\{1 - [(w - 0.5)^2 + 0.75]^p\} \tag{3.38}$$

where ϕ_{b0}, A, D, E are fitting parameters, k_B is the Boltzmann constant, n is the doping level, the positive integer p accounts for the nonlinear dopant kinetics, and ε_S is the permittivity of MoS_2. We have to point out that this is a phenomenological model based on the experimental results reported in (Sangwan et al. 2018). A more general model is needed in neuromorphic computation, since memtransistors on polycrystalline MoS_2 can be grown at the wafer level and very advanced neuromorphic devices could be fabricated based on them.

3.5 Phase-Change Materials (PCMs) Memories

The phase-change materials (PCMs) are dichalcogenides which show a reversible phase transformation from the amorphous state to the crystalline state and vice versa; PCMs are thus resistance-switching memories. The amorphous phase in these materials has a typical high resistance state of hundreds of kiloohms, while the crystalline

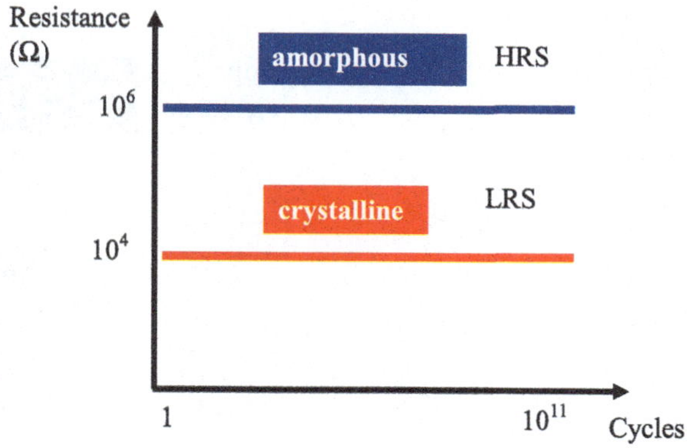

Fig. 3.92 The resistance states of PCMs

state has a resistance of tens of kiloohms. These two states can be maintained about 10^{11} cycles (see Fig. 3.92). The most used PCM material is $Ge_2Sb_2Te_5$ (GST).

The transformation between the states of PCMs is performed by electrical pulses: a low current *set* pulse is used for crystallization of the amorphous phase, while a high current (0.1 mA) *reset* pulse is used to quench the crystalline phase and transform it into an amorphous state; the state is read by a very low current *read* pulse (see Fig. 3.93, where T_{room}, T_{cryst} and T_{melt} denote the room temperature, the crystallization temperature and, respectively, the melting temperature). The electrical pulse provides, via Joule heating, the necessary temperatures for PC transformations (Wouters et al. 2015). The highest temperature necessary to melt the crystalline state is 620 °C in the case of GST. The *set* pulse corresponding to LRS is longer than 100 ns, while the *reset* pulse, corresponding to HRS, is shorter than 70 ns. The PCM memory cell is depicted in Fig. 3.94; the bottom electrode acts as heater and it has an active region where the change of state, from amorphous to polycrystalline, takes place in PCM.

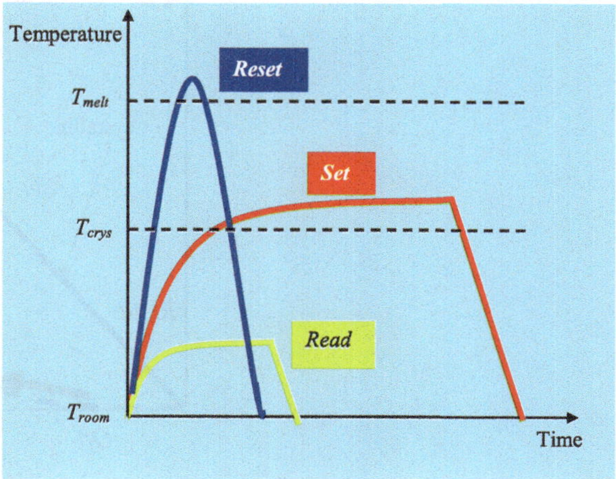

Fig. 3.93 The transformations of PCM via electrical pulses

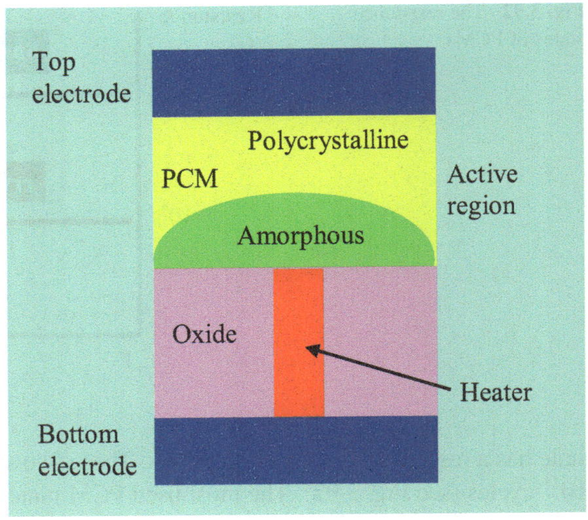

Fig. 3.94 The PCM memory cell

If the above memory cell is connected in series with a load resistance and biased, the current–voltage dependence has the form illustrated in Fig. 3.95. This dependence is S-shaped, showing a negative differential resistance at the *set* onset, and has hysteretic nature. The maximum current is 1 mA and the maximum voltage is 1.5 V, according to Pirovano et al. (2004). In the same reference, a microscopic model showing the role of intrinsic traps is provided.

PCM resistive-switching memories (RSM) technologies are the most advanced among other RSM technologies. The PCM materials are developed since 1970 and are used as storage media in CDs, DVDs, and Blue-Ray disks. It seems that the most

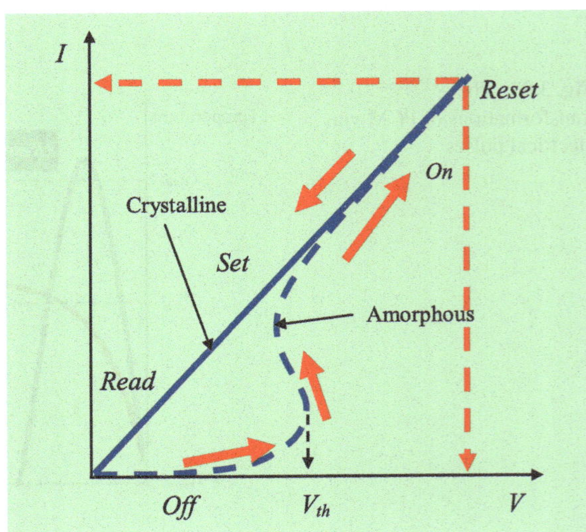

Fig. 3.95 The current-voltage dependence of a PCM memory

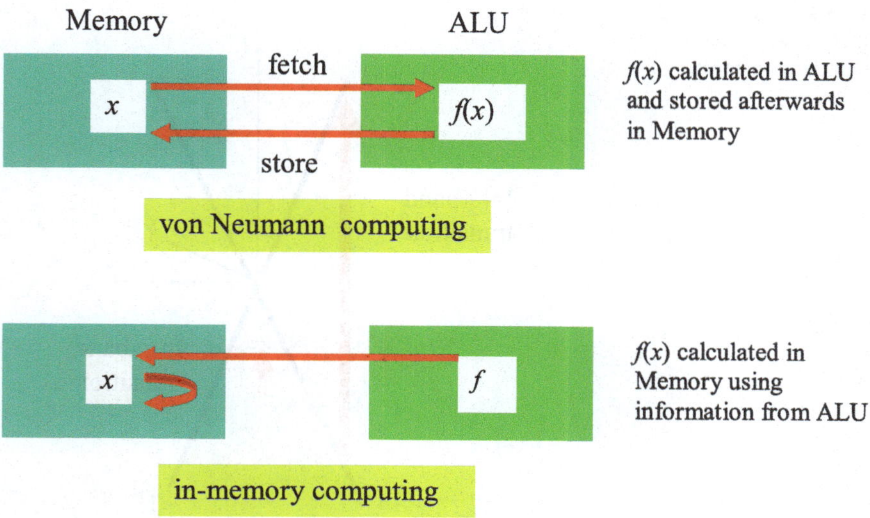

Fig. 3.96 The in-memory computation concept

advanced 3D ReRAM memories to date produced by important companies are based on PCM (Noé et al. 2018).

The most important application of all resistive-switching memories is computing, i.e., in-memory computing (Sebastian et al. 2019). In this case, the necessary information is stored in LRS and HRS of a reversible-state RSM. In a conventional computer, based on the von Neumann architecture, the most energy is consumed in transferring data to and from the memory. In the case of in-memory computing, on the other hand, some computations are performed directly in memory (see Fig. 3.96; ALU stands for Arithmetic Logic Unit), which implies reduced energy consumption.

Logical operations can be performed in RSM and special/dedicated computations can be made. An example of in-memory computing is matrix multiplication, which is a basic mathematical operation in many areas of science. If we want to perform the operation $Ax = b$, the matrix A is encoded in a cross-bar configuration with conductances G_{ij}, the matix x is encoded in voltages applied on the array, while the output currents of the array provide the results of the computation. This issue will be developed in the section dedicated to neuromorphic computing.

3.6 High-Frequency Devices Based on Atomically-Thin Materials

There is an enormous interest in high-frequency devices based on atomically-thin materials due to their applications in communications in high-frequency bands, such as, for example, 5G communications, which are covering the spectrum starting at

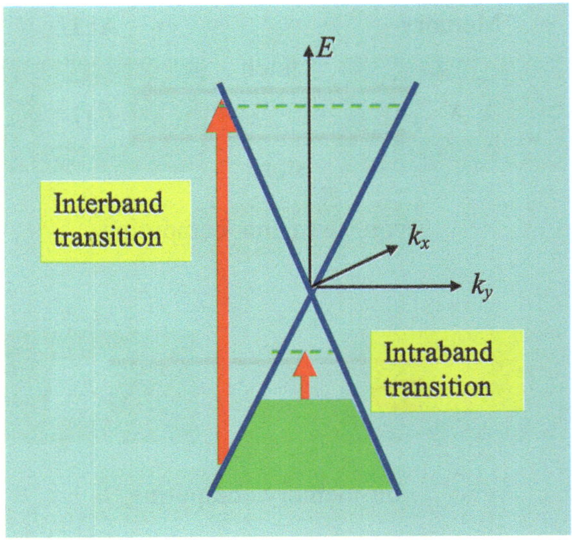

Fig. 3.97 Interband and intraband transitions in graphene monolayer

1 GHz up to 100–300 GHz. Also, a superconducting quantum computer is in fact a complex high-frequency equipment and is developed by all big computing companies. We point out that there are several reviews regarding high-frequency devices in general (Rogalski et al. 2019), as well as reviews on detectors (Rogalski 2019) and high-frequency devices (Dragoman and Dragoman 2016) in 2D materials.

The majority of such high-frequency devices on 2D materials are made on graphene monolayers, where the conductivity plays a key role. For instance, in graphene the charge carrier concentration can be modulated by the gate voltage via electrostatic doping: $n = \alpha V_G$, and the conductivity is determined by interband and intraband transitions in graphene, depicted in Fig. 3.97.

The frequency-dependent conductivity is the sum of frequency-dependent intraband and interband transitions in graphene. According to Hanson (2008):

$$\sigma = \sigma_{intra} + \sigma_{inter} \tag{3.38}$$

where the intraband conductivity is given by

$$\sigma_{\text{intra}} = -i\frac{e^2 k_B T}{\pi \hbar^2(\omega - i2\Gamma)}\left(\frac{\mu_c}{k_B T} + 2\ln[\exp(-\mu_c/k_B T) + 1]\right) \tag{3.39}$$

and the interband conductivity is described by the formula

$$\sigma_{inter} = -i(e^2/4\pi\hbar)\ln\left(\frac{2|\mu_c| - \hbar(\omega - i2\Gamma)}{2|\mu_c| + \hbar(\omega - 2i\Gamma}\right) \tag{3.40}$$

In (3.39)–(3.40), τ is the momentum relaxation time, Γ is a broadening parameter, and μ_c is the electrochemical potential/Fermi energy level in graphene. If $\tau = 40$ fs and we consider small Fermi energy levels, of 100–200 meV, the above formulas tell us that intraband transitions are dominating up to 1 THz, the interband contribution being rather small or negligible. However, if the frequency is greater than few terahertz, the interband transitions are dominating. When $\omega \rightarrow \infty$, the optical conductivity is $e^2/4\hbar$, meaning that 2.3% of the incident light is absorbed in graphene at normal incidence, independent of the wavelength of the electromagnetic radiation.

In the terahertz frequency range, both types of conductivity, i.e., interband and intraband, must be considered to contribute to the surface impedance, which is written as

$$Z_S(V_G) = 1/\sigma(\omega) = R_s(V_G) + jX_s(V_G). \tag{3.41}$$

Simulations demonstrate that the surface impedance of a graphene monolayer is reactive (inductive) in the THz range (Dragoman et al. 2014b), R_s having values less than 52 Ω, but $X_s \gg R_s$ and has values in the range of 17–20 kΩ. Graphene as reactive impedance surface (RIS) has several advantages as a substrate in antennas: (i) reduction of the lateral lobes, (ii) improvement of the front-to-back ratio of radiated power, and (iii) miniaturization. Since Z_S in (3.41) is dependent on the Fermi level via the applied gate voltage, which influences also the concentration of charge carriers, graphene in terahertz is a tunable RIS.

For example, let us consider a typical terahertz antenna on graphene, such as a gold bow-tie antennas, with dimensions $L = 0.38\lambda_0 = 11.25$ μm, $W = 1$ μm, a gap of 100 nm, and gold thickness of 100 nm (see Fig. 3.98). The real and imaginary parts of the input impedance is displayed in Fig. 3.99.

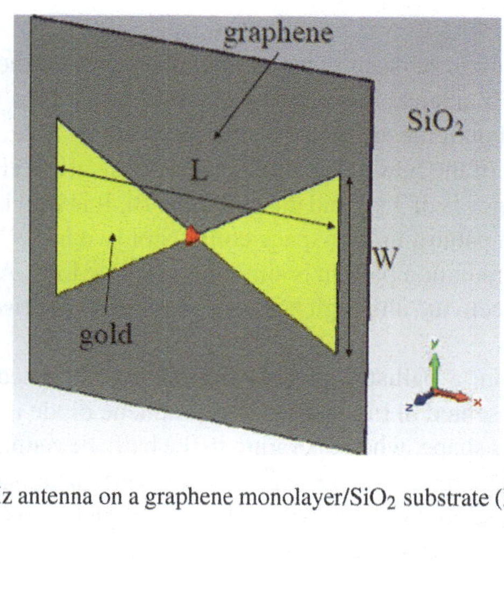

Fig. 3.98 Bow-tie THz antenna on a graphene monolayer/SiO_2 substrate (Dragoman et al. 2014b)

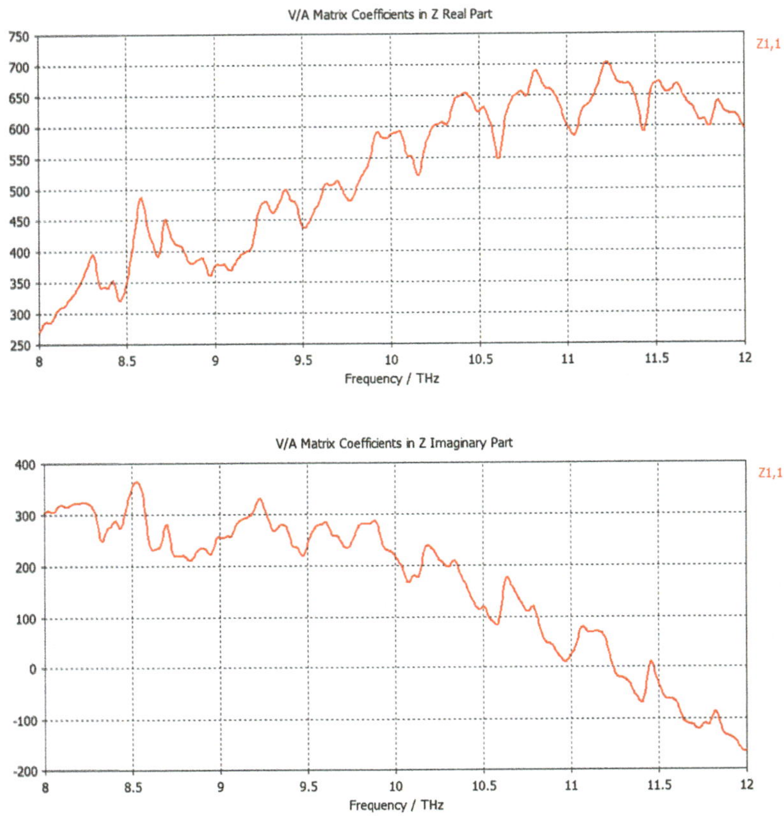

(a)

(b)

Fig. 3.99 Real and imaginary parts of the input impedance in a THz bow-tie antenna on graphene (Dragoman et al. 2014b)

From Fig. 3.99 it can be seen that the imaginary part of the impedance of the graphene bow-tie THz antenna is becoming zero beyond 11 THz, and, due to the high resistance of the real part, the antenna has efficiency below 50%. For example, from the radiation pattern of the bow-tie antenna on graphene at 10 THz (see Fig. 3.100), the simulated efficiency is of 33% and the gain is 8 dBi. It is well known that the bow-tie antenna radiation pattern in free-space consists of two lobes, but since graphene is a RIS in THz, the radiation pattern is similar to a single lobe. As such, the antenna has a pretty good directivity, although it is very small in comparison to the free-space wavelength.

For THz detection, a ballistic geometric diode can be used (Dragoman et al. 2014b), as that represented in Fig. 3.101. This graphene diode is rectifying the electrical signal due to its shape, when operating in the ballistic regime at room temperature (Dragoman and Dragoman 2013). No *p-n* junction is necessary for rectification, but the length of the diode must be below 400 nm, which is the mean-free-path of carriers in graphene on Si/SiO_2 substrates.

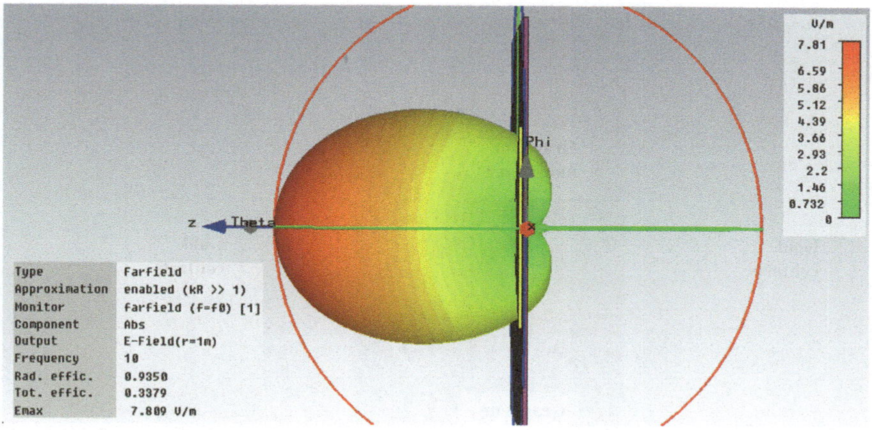

Fig. 3.100 Radiation pattern of the graphene bow-tie THz antenna (Dragoman et al. 2014b)

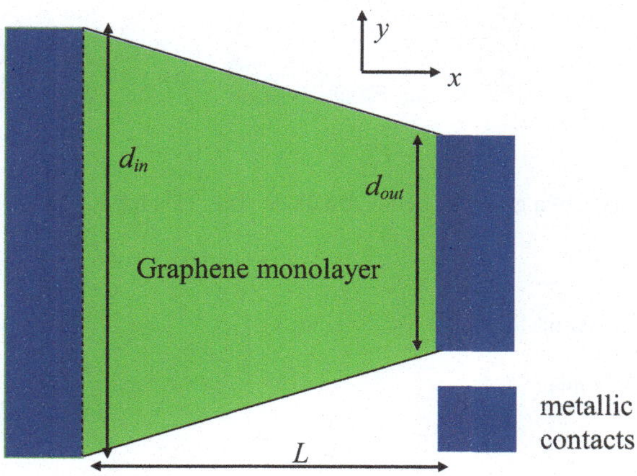

Fig. 3.101 Ballistic geometric diode

The geometric diode in Fig. 3.101 was fabricated on a graphene monolayer deposited on a 4 in. wafer. The graphene monolayer is grown by CVD and is deposited on 285 nm of SiO_2 grown on *p*-doped Si. The SEM photo of the diode is presented below, in Fig. 3.102. The graphene diode has a length of 100 nm, a shoulder of d_{in} = 100 nm and a neck of only $d_{\text{out}} = 30$ nm.

The diode has a zero region of about 150 meV around 0 V, where the current is very small, below 0.1 nA, while beyond this region the current has an almost exponential behavior, reaching ±10 μA at voltages of ±1 V. The current–voltage dependence can be tuned by applying back-gate voltages.

The responsivity of this THz detector formed from a bow-tie antenna, which has in its arm the geometric diode, is displayed in Fig. 3.103. It can be seen that

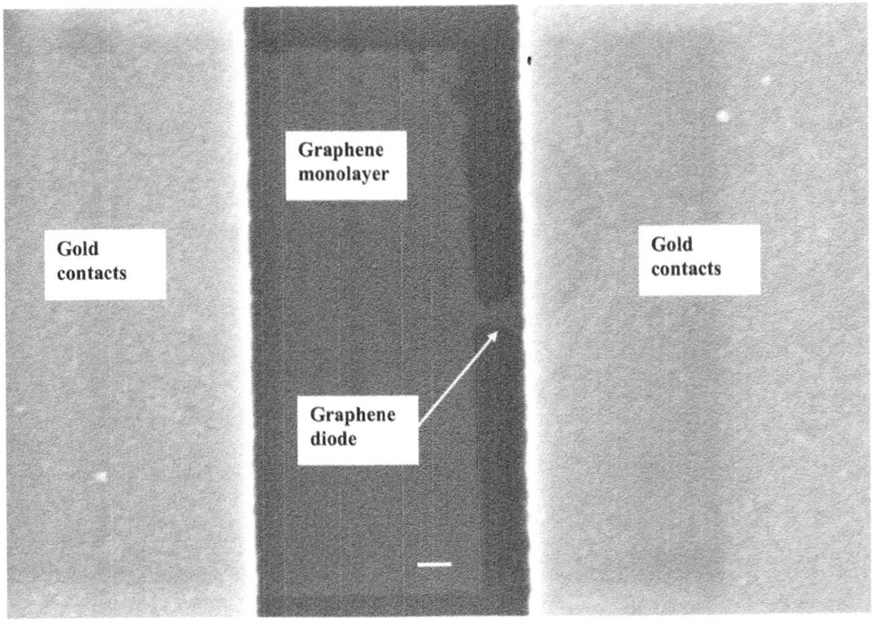

Fig. 3.102 SEM image of a graphene geometric diode. The scale bar is 100 nm (Dragoman et al. 2014a)

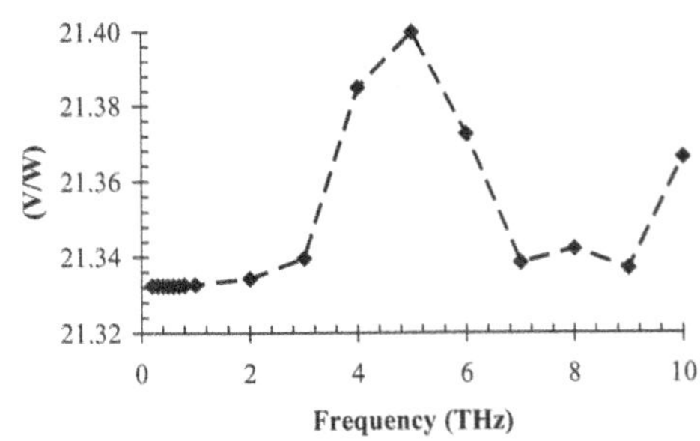

Fig. 3.103 Responsivity of the bow-tie loaded with a geometric diode (Dragoman et al. 2014b)

the responsivity is almost constant, of 21.33 V/W, having a small relative peak of 21.4 V/W at 5 THz.

The detection of high-frequency signals up to the THz band can be performed also with transistors based on graphene and other 2D materials. The detection of such high-frequency signals with amplitude V_a relies on the nonlinear dependences of the drain current versus gate voltage ($I_D - V_G$), and the detected signal has the expression (Zak et al. 2014):

$$\Delta U_{\mathrm{det}} = (V_a^2/4\sigma)(\mathrm{d}\sigma/\mathrm{d}V_G) \tag{3.42}$$

where

$$\sigma = (L/W)\sigma_{DS} \tag{3.43}$$

where σ_{DS} is the drain-source conductivity. Its nonlinear dependence on the gate voltage is the physical basis of the THz detection with the help of FETs based on 2D materials.

For detection, the THz signal is applied between gate and source and the detected signal is collected between drain and source. The gate and source electrodes have the shape of an antenna, which collect the THz signal. Very often, a Si lens is used between the antenna arms to focus the signal between gate and source electrodes. Figure 3.104 shows a schematic THz detector at room temperature using a graphene FET transistor.

Room temperature detectors were fabricated using various types of antennas and graphene monolayers or bilayers. In this respect, a 0.3 THz signal at room temperature has been detected (Vicarelli et al. 2012) using a log-periodic circular-toothed antenna with an outer radius of 322 μm, the two arms of which acting at the same time as source and gate electrodes of a graphene FET having as channel a graphene bilayer. The channel length in this case is of 10 μm and the gate length is 200 nm.

In another example, a splitting gate configuration, and thus a split bow-tie antenna, was used to detect a 0.6 THz signal at room temperature (Zak et al. 2014). The graphene FET in this situation has a channel length of 2.5 μm and a gate width of

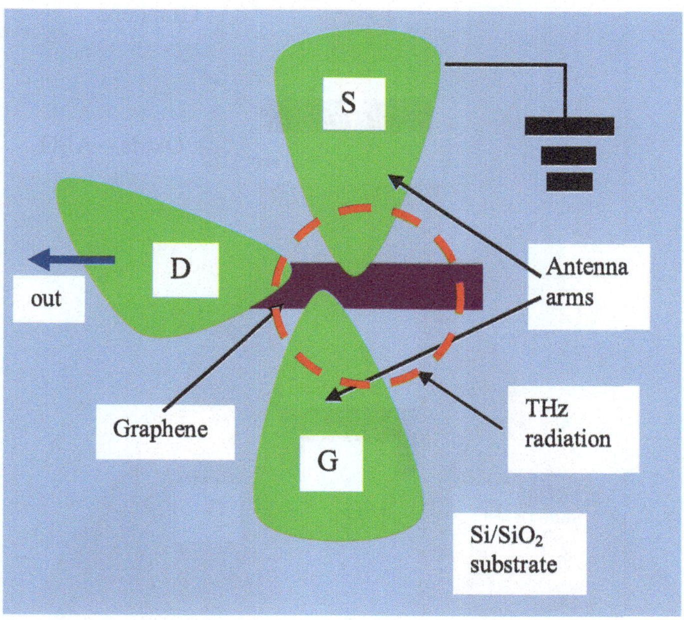

Fig. 3.104 Room temperature THz detector principle using a graphene FET

2 μm. The results have shown a noise-equivalent-power (NEP) of 500 $pW/Hz^{1/2}$ and a responsivity of 14 V/W. The split bow-tie antenna is represented schematically in Fig. 3.105.

In yet another example, very good results for THz signal detection are obtained if a bridge formed by self-switching diodes (SSD) is used (Zhang et al. 2019). The concept of SSD was discussed earlier in this chapter.

THz detection at room temperature is possible also using the photothermoelectric effect in graphene (Cai et al. 2014). Since graphene has an interband absorption in the THz range, the heat capacity in graphene is much smaller than in bulk materials. This physical property means that a significant temperature gradient $\nabla T(x)$ can be produced in few picoseconds. In the experiment of Cai et al. (2014), two dissimilar metals play the role of electrodes and are deposited over a graphene monolayer (see Fig. 3.106); the asymmetry of metal contacts generates a net current due to the thermoelectric effect. The distance between the metal contacts is 3 μm. The dissimilar metals are Cr and Au. This is not an easy metal deposition process, since these metal combinations are obtained using consecutive and different angle evaporations. The

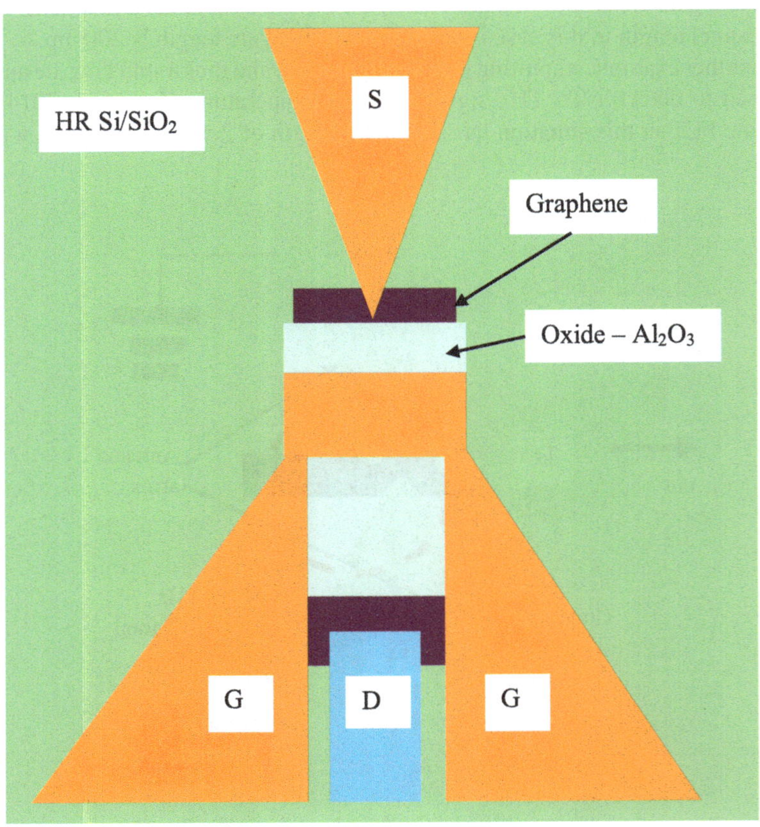

Fig. 3.105 The split bow-tie THz detector based on a graphene FET

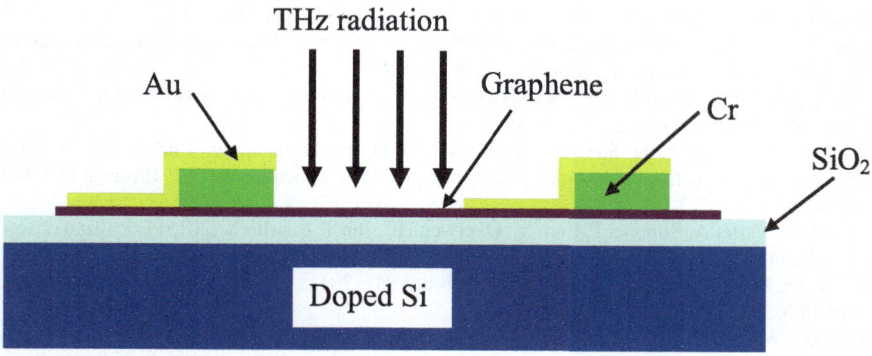

Fig. 3.106 Photothermoelectric THz detection

carriers in graphene are heated by the incoming THz radiation, and the diffusion of charge carriers generate the potential gradient (Cai et al. 2014)

$$\nabla V(x) = -S\nabla T(x), \tag{3.44}$$

where S is the Seebeck coefficient of graphene. In this context, note that the Seebeck coefficient can reach the huge value of 30 mV/K in a periodically gated graphene monolayer structure (Dragoman and Dragoman 2007b).

The THz detector response is the directly proportional with the potential generated by the temperature gradient:

$$v = \int_0^L \nabla V(x)\mathrm{d}x \tag{21}$$

The photothermoelectric THz detector has a responsivity of 10 V/W and a NEP of 1100 pW $Hz^{-1/2}$ if calculated with respect to the incident power, and of 700 V/W and, respectively, 20 pW $Hz^{-1/2}$ if calculated relative to the absorbed power.

References

Aharonovich I, Englund D, Milos Toth M (2016) Solid-state single-photon emitters. Nat Photonics 1:631–641

Alam AA, Si M, Ye PD (2019) A critical review of recent progress on negative capacitance transistors. Appl Phys Lett 114:090401

Auton G, Zhang J, Kumar RK, Wang H, Zhang X, Wang Q, Hill E, Song A (2016) Graphene ballistic nano-rectifier with very high responsivity. Nat Commun 7:11670

Auton G, But DB, Zhang J, Hill E, Coquillat D, Consejo C, Nouvel P, Knap W, Varani L, Teppe F, Torres J, Song A (2017) Terahertz detection and imaging using graphene ballistic rectifiers. Nano Lett 17:7015

Bain AK, Chand P (2017) Ferroelectrics: principles and applications. Wiley-VCH
Barreiro A, van der Zant HSJ, Vandersypen LMK (2012) Quantum dots at room temperature carved out from few-layer graphene. Nano Lett 12:6096–6100
Bertolazzi S, Krasnozhon D, Kiss A (2013) Nonvolatile memory cells based on MoS_2/graphene heterostructures. ACS Nano 7:3246–3252
Bidenko P, Lee S, Han J-H, Song J-D, Kim S-H (2018) Simulation study of the design sub-kT/q non-hysteretic negative capacitance FET using capacitive matching. J IEEE Electron Dev Soc 6:901–923
Bischoff D, Varlet A, Simonet P, Eich M, Overweg HC, Ihn T, Ensslin K (2015) Localized charge carriers in graphene nanodevices. Appl Phys Rev 2:031301
Böscke TS, Müller J, Bräuhaus D, Schröder U, Böttger U (2011) Ferroelectricity in hafnium oxide thin films. Appl Phys Lett 99:102903
Brady GJ, Way AJ, Safron NS, Evensen HT, Gopalan P, Arnold MS (2016) Quasi-ballistic carbon nanotube array transistors with current density exceeding Si and GaAs. Sci Adv 2:e1601240
Cai X, Sushkov AB, Suess RJ, Jadidi MM, Jenkins GJ, Nyakiti LO, Meyers-Ward RL, Li S, Yan J, Gaskill DK, Murphy TE, Drew HD, Fuhrer MS (2014) Sensitive room-temperature terahertz detection via the photothermoelectric effect in graphene. Nat Nanotechnol 9:814–819
Cheng R, Jiang S, Chen Y, Liu Y, Weiss N, Cheng H-C, Wu H, Huang Y, Duan X (2014) Few-layer molybdenum disulfide transistors and circuits for high-speed flexible electronics. Nat Commun 5:5143
Cheng C-H, Fan C-C, Tu C-Y, Hsu H-H, Chang C-Y (2019) Implementation of dopant-free hafnium oxide negative capacitance field-effect transistor. IEEE Trans Electron Dev Lett 66:825–828
Chua L (2014a) If it's pinched it's a memristor. In: Tetzlaff R (ed) Memristors and memristive systems. Springer, pp 17–90
Chua L (2014b) Memristor, Hodgkin-Huxley, and edge of chaos. In: Adamatzky A, Chua L (eds) Memristor networks. Springer, pp 67–94
Cui C, Xue F, Hu W-J, Li L-J (2018) Two-dimensional materials with piezoelectric and ferroelectric functionalities. NPJ 2D Mater Appl 18:18
Datta S (1997) Electronic transport in mesoscopic systems. Cambridge University Press
Desai SB, Madhvapathy SR, Sachid AB, Llinas JP, Wang Q, Ahn GH, Pitner G, Kim MJ, Bokor J, Hu C, Wong HP, Javey A (2016) MoS_2 transistors with 1-nanometer gate lengths. Science 354:99–102
Diduck Q, Irie H, Margala M (2009) A room temperature ballistic deflection transistor for high performance applications. Int J High Speed Electron Syst 19:23–31
Dragoman D, Dragoman M (2007a) Negative differential resistance of electrons in graphene barrier. Appl Phys Lett 90:143111
Dragoman D, Dragoman M (2007b) Giant thermoelectric effect in graphene. Appl Phys Lett 91:203116
Dragoman M, Dragoman D (2009) Nanoelectronics: principles and devices (2nd edn). Artech House
Dragoman D, Dragoman M (2013) Geometrically induced rectification in two-dimensional ballistic nanodevices. J Phys D 46:055306
Dragoman M, Dragoman D (2016) THz devices based on carbon nanomaterials. In: Tiginyanu I, Topala P, Ursaki V (eds) Nanostructures and thin films for multifunctional applications. Nanoscience and technology series. Springer, pp 533–549
Dragoman M, Dragoman D (2017) 2D nanoelectronics, physics and devices of atomically thin materials. Springer, Berlin
Dragoman M, Dinescu A, Dragoman D (2014a) Negative differential resistance in graphene-based ballistic field-effect-transistor with oblique top gate. Nanotechnology 25:41520
Dragoman M, Aldrigo M, Dinescu A, Dragoman D, Costanzo A (2014b) Towards a terahertz direct receiver based on graphene up to 10 THz. J Appl Phys 115:044307
Dragoman M, Dinescu A, Dragoman D (2016a) Room temperature on-wafer ballistic graphene field-effect-transistor with oblique double-gate. J Appl Phys 119:244305

Dragoman M, Tiginyanu I, Dragoman D, Braniste T, Ciobanu V (2016b) Memristive GaN ultrathin suspended membrane array. Nanotechnology 27:295204

Dragoman M, Dinescu A, Dragoman D (2017) Room temperature nanostructured graphene transistor with high on/off ratio. Nanotechnology 28:015201

Dragoman M, Dinescu A, Dragoman D (2018a) Wafer-scale fabrication and room-temperature experiments on graphene-based gates for quantum computation. IEEE Trans Nanotechnol 17:362–366

Dragoman M, Dinescu A, Dragoman D (2018b) Solving the graphene electronics conundrum: high mobility and high on-off ratio in graphene nanopatterned transistors. Physica E 97:296–301

Dragoman M, Modreanu M, Povey IM, Dinescu A, Dragoman D, Di Donato A, Pavoni E, Farina M (2018c) Wafer-scale very large memory windows in graphene monolayer/HfZrO ferroelectric capacitors. Nanotechnology 29:425204

Dragoman M, Tiginyanu I, Dragoman D, Dinescu A, Braniste T, Ciobanu V (2018d) Learning mechanisms in memristor networks based on GaN nanomembranes. J Appl Phys 124:152110

Dragoman M, Batiri M, Dinescu A, Ciobanu V, Rusu E, Dragoman D, Tiginyanu I (2018e) A SnS_2-based photomemristor driven by sun. J Appl Phys 123:024506

Dragoman M, Dinescu A, Dragoman D (2019) 2D materials nanoelectronics: new concepts, fabrication, characterization from microwaves up to optical spectrum. Phys Status Solidi A 216:1800724

Dragoman M, Aldrigo M, Connolly J, Povey IM, Iordanescu S, Dinescu A, Vasilache D, Modreanu M (2020a) MoS_2 radio: detecting radio-waves with a two-dimensional (2D) transition metal dichalcogenide semiconductor. Nanotechnology 31:06LT01

Dragoman M, Modreanu M, Povey IM, Dinescu A, Dragoman D (2020b) Reconfigurable horizontal-vertical carrier transport in graphene/HfZrO field-effect-transistor. Nanotechnology 31:025203

Ferry DF, Goodnick M (1997) Transport in nanostructures. Cambridge University Press

Gehring P, Sadeghi H, Sangtarash S, Lau CS, Liu J, Ardavan A, Warner JH, Lambert CJ, Briggs GA, Mol JA (2016) Quantum interference in graphene nanoconstrictions. Nano Lett 16:4215–4216

Georgiou T, Jalil R, Belle BD, Britnell L, Gorbachev RV, Morozov SV, Kim Y-J, Gholinia A, Haig SJ, Makarovsky O, Eaves L, Ponomarenko LA, Geim AK, Novoselov KS, Mishchenko A (2012) Vertical field-effect transistor based on graphene WS_2 heterostructures for flexible and transparent electronics. Nat Nanotechnol 8:100–103

Giovannetti G, Khomyakov PA, Brocks G, Kelly PJ, van den Brink J (2007) Substrate-induced band gap in graphene on hexagonal boron nitride: Ab initio density functional calculations. Phys Rev B 76:073103

Gonzalo JA, Jiménez B (2007) Ferroelectricity: the fundamentals collection. Wiley VCH

Güttinger J, Molitor F, Stampfer C, Schnez S, Jacobsen A, Dröscher S, Ihn T, Ensslin K (2012) Transport through graphene quantum dots. Rep Prog Phys 75:126502

Hammam AMM, Schmidt ME, Muruganathan M, Suzuki S, Mizuta H (2018) Sub-10 nm graphene nano-ribbon tunnel field-effect transistor. Carbon 126:588–593

Han S-J, Valdes Garcia A, Oida S, Jenkins KA, Haensch W (2014) Graphene radio frequency receiver integrated circuit. Nat Commun 5:3086

Hanson GW (2008) Dyadic Green's functions and guided surface waves for a surface conductivity model of graphene. J Appl Phys 103:064302

Harrison P, Valavanis A (2016) Quantum wells, wires and dots (4th ed). Wiley

Hu SG, Wu SY, Jia WW, Yu Q, Deng LJ, Fu YQ, Liu Y, Chen TP (2014) Review of nanostructured resistive switching memristor and its applications. Nanosci Nanotechnol Lett 6:729–757

Ilatikhameneh H, Ameen T, Novakovic B, Tan Y, Klimeck G, Rahman R (2016) Saving Moore's law down to 1 nm channels with anisotropic effective mass. Sci Rep 6:31501

Ionescu AM, Riel H (2011) Tunnel field-effect transistors as energy-efficient electronic switches. Nature 479:329–337

Jadwiszczak J, Keane D, Maguire P, Cullen CP, Zhou Y, Song H, Downing C, Fox D, McEvoy N, Zhu R, Xu J, Duesberg GS, Liao Z-M, Boland JJ, Zhang H (2019) MoS_2 memtransistors fabricated by localized helium ion beam irradiation. ACS Nano 13:14262–14273

Ji H, Wei J, Natelson D (2012) Modulation of the electrical properties of VO_2 nanobeams using an ionic liquid as a gating medium. Nano Lett 12:2988

Jie W, Hao J (2014) Graphene-based hybrid structures combined with functional materials of ferroelectrics and semiconductors. Nanoscale 6:6346–6362

Jo SH, Chang T, Ebong I, Bhadviya BB, Mazumder P, Lu W (2010) Nanoscale memristor device as synapse in neuromorphic systems. Nano Lett 10:1297–1301

Kang K, Xie S, Huang L, Han Y, Huang PY, Mak KF, Kim C-J, Muller D, Park J (2015) High-mobility three-atom-thick semiconducting films with wafer-scale homogeneity. Nature 520:656–660

Kim M, Ge R, Wu X, Lan X, Tice J, Lee JC, Akinwande D (2018) Zero-static power radio-frequency switches based on MoS_2 atomristors. Nat Commun 9:2524

Knap W, Teppe F, Dyakonova N, Coquillat D, Łusakowski J (2008) Plasma wave oscillations in nanometer field effect transistors for terahertz detection and emission. J Phys Condens Matter 20:384205

Krivokapic Z, Rana U, Galatage R, Razavieh A, Aziz A, Liu J, Shi J, Kim HJ, Sporer R, Serrao C, Busquet A, Polakowski P, Müller J, Kleemeier W, Jacob A, Brown D, Knorr A, Carter R, Banna S (2017) 14 nm ferroelectric FinFET technology with steep subthreshold slope for ultra low power applications. IEEE Int Electron Devices Meet (IEDM):15.1.1–15.1.4

Kwon D-H, Kim KM, Jang JH, Jeon JM, Lee MH, Kim GH, Li X-S, Park G-S, Lee B, Han S, Kim M, Hwang CS (2010) Atomic structure of conducting nanofilaments in TiO_2 resistive switching memory. Nat Nanotechnol 5:148–153

Lee G-H, Kim S, Jhi S-H, Lee H-J (2015) Ultimately short ballistic vertical graphene Josephson junctions. Nat Commun 6:6181

Li T, Gallop J, Hao L, Romans E (2018) Ballistic Josephson junctions based on CVD graphene. Supercond Sci Technol 31:045004

Liu XL, Hug D, Vandersypen LVK (2010) Gate-defined graphene double quantum dot and excited state spectroscopy. Nano Lett 10:1623–1627

Liu Y, Duan X, Huang Y, Duan X (2018) Two-dimensional transistors beyond graphene and TMDCs. Chem Soc Rev 47:6388

Lusakowski J, Knap W, Dyakonova N, Varani L, Mateos J, Gonzalez T, Roelens Y, Bollaert S, Cappy A (2005) Voltage tuneable terahertz emission from a ballistic nanometer InGaAs/InAlAs transistor. J Appl Phys 97:064307

Lv Y, Qin W, Wang C, Liao L, Liu X (2019) Recent advances in low-dimensional heterojunction-based tunnel field effect transistors. Adv Electron Matt 5:1800569

McGuire FA, Lin YC, Price K, Rayner GB, Khandelwal S, Salahuddin S, Franklin AD (2017) Sustained sub-60 mV/decade switching via the negative capacitance effect in MoS_2 transistors. Nano Lett 17:4801–4806

Michel P (ed) (2017) Quantum dots for quantum information technologies. Springer

Moon JS, Curtis D, Bui S, Hu M, Gaskill DK, Tedesco JL, Asbeck P, Jernigan GG, VanMil BL, Myers-Ward RL, Eddy CR Jr, Campbell PM, Weng X (2010) Top-gated epitaxial graphene FETs on Si-face SiC wafers with a peak transconductance of 600 mS/mm. IEEE Electron Dev Lett 31:260–262

Moore GE (1965) Cramming more components onto integrated circuits. Electronics 38:114

Mott NF (1949) The basis of the electron theory of metals, with special reference to the transition metals. Proc Phys Soc A 62:416

Müller J, Böscke TS, Schröder U, Mueller S, Bräuhaus D, Böttger U, Frey L, Mikolajick T (2012) Ferroelectricity in simple binary ZrO_2 and HfO_2. Nano Lett 12:4318–4323

Nakano M, Shibuya K, Okuyama D, Hatano T, Ono S, Kawasaki M, Iwasa Y, Tokura Y (2012) Collective bulk carrier delocalization driven by electrostatic surface charge accumulation. Nature 487:459–462

Noé P, Vallée C, Hippert F, Fillot F, Raty J-Y (2018) Phase-change materials for non-volatile memory devices: from technological challenges to materials science issues. Semicond Sci Technol 33:013002

Nemnes G A,Dragoman D, Dragoman M (2019) Graphene bandgap induced by ferroelectric Pca21 HfO2 substrate: a first-principles study. Phys.Chem. Phys: 21,15001

Pan Y, Yin H, Huang K, Zhang Z, Zhang Q, Jia K, Wu Z, Luo K, Yu J, Li J, Wang W, Ye T (2019) Novel 10-nm gate length MoS_2 transistor fabricated on Si fin substrate. IEEE J Electron Dev Soc 7:484–488

Park MH, Lee YH, Kim HJ, Kim YJ, Moon T, Kim KD, Muller J, Kersch A, Schroeder U, Mikolajick T, Hwang CS (2015) Ferroelectricity and antiferroelectricity of doped thin HfO_2-based films. Adv Mater 27:1811–1831

Pergament A, Crunteanu A, Beaumont A, Stefanovich G (2015) Vanadium dioxide: metal-insulator transition, electrical switching and oscillations. A review of state of the art and recent progress. EMN meeting on computation and theory, energy materials and nanotechnology, November 9 to 12, Istanbul, Turkey. https://arxiv.org/abs/1601.06246

Pirovano A, Lacaita AL, Pellizzer F, Kostylev SA, Benvenuti A, Bez R (2004) Low-field amorphous state resistance and threshold voltage drift in chalcogenide materials. IEEE Trans Electron Dev 51:714–719

Rogalski A (2019) Graphene-based materials in the infrared and terahertz detector families: a tutorial. Adv Opt Photonics 11:314–379

Rogalski A, Kopytko M, Martyniuk P (2019) Two-dimensional infrared and terahertz detectors: outlook and status. Appl Phys Rev 6:021316

S. Salahuddin , S. Datta (2008) Use of negative capacitance to provide voltage amplification for low power nanoscale devices. Nano Lett. 8: 405-410

Sandner A, Preis T, Schell C, Giudici P, Watanabe K, Taniguchi T, Weiss D, Eroms J (2015) Ballistic transport in graphene antidot lattices. Nano Lett 15:8402–8406

Sangwan VK, Lee H-S, Bergeron H, Balla I, Beck ME, Chen K-S, Hersam MC (2018) Multi-terminal memtransistors from polycrystalline monolayer molybdenum disulfide. Nature 554:500–504

Sanne A, Ghosh R, Rai A, Yogeesh MN, Shin SH, Sharma A, Jarvis K, Mathew L, Rao R, Akinwande D, Banerjee S (2015) Radio frequency transistors and circuits based on CVD MoS_2. Nano Lett 15:5039

Schmidt FE, Jenkins MD, Watanabe K, Steele GA (2018) A ballistic graphene superconducting microwave circuit. Nat Commun 9:4

Schröder U, Mueller S, Mueller J, Yarchuk E, Martin D, Adelmann C, Schloesser T, van Bentum R, Mikolajick T (2013) Hafnium oxide based CMOS compatible ferroelectric materials. ECS J Solid State Sci Technol 2:N69–N72

Sebastian A, Le Gallo M, Eleftheriou E (2019) Computational phase-change memory: beyond von Neumann computing. J Phys D 52:443002

Sengupta S, Wang SK, Liu K, Bhat K, Dhara S, Wu J, Deshmukh MM (2011) Field-effect modulation of conductance in VO_2 nanobeam transistors with HfO_2 as the gate dielectric. Appl Phys Lett 99:062114

Shao Z, Cao X, Luo H, Jin P (2018) Recent progress in phase-transition mechanism and modulation of vanadium dioxide materials. NPG Asia Mater 10:581–605

Sharma P, Zhang J, Ni K, Datta S (2018) Time-resolved measurement of negative capacitance. IEEE Electron Dev Lett 39:272–275

Shukla N, Thathachary AV, Agrawal A, Paik H, Aziz A, Schlom DG, Gupta SK, Engel-Herbert R, Datta S (2015) A steep-slope transistor based on abrupt electronic phase transition. Nat Commun 6:7812

Si M, Su CJ, Jiang C, Conrad NJ, Zhou H, Maize KD, Qiu G, Wu C-T, Shakouri A, Alam MA, Ye PD (2018) Steep-slope hysteresis-free negative capacitance MoS_2 transistors. Nat Nanotechnol 13:24–28

Slesazeck S, Mikolajick Thomas (2019) Nanoscale resistive switching memory devices: a review. Nanotechnology 30:352003

Song AM (2004) Room-temperature ballistic nanodevices. In: Nalwa HS (ed) Encyclopedia of nanoscience and nanotechnology (vol 9). American Scientific Publisher, pp 371–389

Strukov DB, Snider GS, Stewart DR, Williams RS (2008) The missing memristor found. Nature 453:80–83
Tetzlaff R (2014) Memristors and memristive systems. Springer
Tran T, Bray K, Ford MJ, Toth M, Aharonovich I (2016) Quantum emission from hexagonal boron nitride monolayers. Nat Nanotechnol 11:37–41
Vicarelli L, Vitiello MS, Coquillat D, Lombardo A, Ferrari AC, Knap W, Polini M, Pellegrini V, Tredicucci A (2012) Graphene field-effect transistors as room-temperature terahertz detectors. Nat Mater 11:865–871
Wachter S, Polyushkin D, Bethge K, Mueller T (2017) A microprocessor based on a two-dimensional semiconductor. Nat Commun 8:14948
Wang ZM (ed) (2012) Quantum Dot devices, Springer
Wang M, Cai S, Pan C, Wang C, Lian X, Zhuo Y, Xu K, Cao T, Pan X, Wang B, Liang S-J, Yang JJ, Wang P, Miao F (2018) Robust memristors based on layered two-dimensional materials. Nat Electron 1:130–136
Wang L, Liao W, Wong SL, Yu ZG, Li S, Lim Y-F, Feng X, Tan WC, Huang X, Chen L, Liu L, Chen J, Gong X, Zhu C, Liu X, Zhang Y-W, Chi D, Ang K-W (2019a) Artificial synapses based on multiterminal memtransistors for neuromorphic application. Adv Func Mater 29:1901106
Wang X, Yu P, Lei Z, Zhu C, Cao X, Liu F, You L, Zeng Q, Deng Y, Zhu C, Zhou J, Fu Q, Wang J, Huang Y, Liu Z (2019b) Van der Waals negative capacitance transistors. Nat Commun 10:3037
Wong JC, Salahuddin S (2019) Negative capacitance transistors. Proc IEEE 107:49–62
Wouters DJ, Waser R, Wuttig M (2015) Phase-change and redox-based resistive switching memories. Proc IEEE 103:1274–1288
Ye P, Ernst T, Khare MV (2019) The last silicon transistor: nanosheet devices could be the final evolutionary step for Moore's Law. IEEE Spectr 56:30–35
Zak A, Andersson MA, Bauer M, Matukas J, Lisauskas A, Roskos HG, Stake J (2014) Antenna-integrated 0.6 THz FET direct detectors based on CVD graphene. Nano Lett 14:5834–5838
Zhang L, Chan M (2016) Tunneling field effect transistor technology. Spinger Nature
Zhang X, Grajal J, Wang X, Radhakrishna U, Zhang Y, Kong J, Dresselhaus MS, Palacios T (2018) MoS_2 phase-junction-based Schottky diodes for RF electronics. IEEE/MTT-S international microwave symposium, Philadelphia, USA, p 345
Zhang L, Gong T, Wang H, Zhinan G, Zhang H (2019) Memristive devices based on emerging two-dimensional materials beyond graphene. Nanoscale 11:12413–12435
Zhou C, Chai Y (2017) Ferroelectric-gated two-dimensional materials-based electronic devices. Adv Electron Mater 3:1600400
Zhou Y, Ramanathan S (2015) Mott memory and neuromorphic devices. Proc IEEE 103:1289–1310
Zhou Y, Gweon G-H, Fedorov AV, First PN, de Heer WA, Lee D-H, Guinea F, Castro Neto AH, Lanzara A (2007) Substrate-induced bandgap opening in epitaxial graphene. Nat Mater 6:770–775

Chapter 4
Quantum Computing

Abstract In this chapter we present the physical basis of quantum computing and provide examples linked to electronics at atomic scale, i.e., quantum computing systems based on superconducting Josephson junctions and quantum dots. The entire chapter is focused on applications, and only the basic theory is presented, since the majority of the papers dedicated to quantum computing are theoretical papers. The basic knowledge from previous chapters, such as tunneling effects, quantum dots, SETs and FETs are needed to fully understand what a quantum processor is.

4.1 The Physical Basis of Quantum Computation

4.1.1 Bits and the Qubits

It is well-known that the computers used by us every day are based on the fact that all computations are made using 0 and 1 bits. Why? The reason is that all logical operations made on bits stored in registers (series of bits) can be performed using only two logical operations: AND and NOT, described below in Fig. 4.1. Any other logical operation on the 0 and 1 bits can be decomposed in a series of AND and NOT operations, which in computers are implemented by logical gates based on transistors.

What is the physical nature of a bit? The 0 and 1 states can be, for instance, the two states of any switch, termed in the previous chapters as *on* and *off* states. A transistor has two distinctive states. The *on* state corresponds to the saturation region of the $I_D - V_D$ dependence at a certain gate voltage, such that the drain current has a constant, high value, irrespective of the drain voltage. The *off* state, on the other hand, appears at another gate voltage value, for which the drain current should be zero, ideally, but in reality it has a small value, called leakage current. The ratio between currents in the *on* and *off* state is 10^3. The computing performances depend entirely on the number of transistors able to perform complex tasks and on their switching time. Today, a microprocessor contains tens of billions of transistors as a result of continuous reduction of the dimensions of the transistors during the last fifty years.

M. Dragoman and D. Dragoman, *Atomic-Scale Electronics Beyond CMOS*,
https://doi.org/10.1007/978-3-030-60563-6_4

A	*B*	*A AND B*
0	0	0
1	0	0
0	1	0
1	1	1

A	*NOT A*
0	1
1	0

Fig. 4.1 Truth tables (top) and representations (bottom) of the AND and NOT logical gates

The most important enemies of this continuous reduction of transistor dimensions on chip are the short channel effects, due to quantum mechanical phenomena, and the limits of optical lithography. However, the most adverse effect of such dense chips is the dissipated heat generated by billions of transistors connected with tens of billions of metallic wires. This dissipated heat is so big that it cannot be evacuated with simple cooling devices. Therefore, by using simple cooling devices for microprocessors, their speed is internationally limited at few gigahertz and is stalled at these value since 20 years. Today, each microprocessor has a couple of cores, which, paradoxically, are less used every generation. The dramatic effect is that some core areas are powered-off many times during microprocessor operations, and effect termed as "dark silicon" (Esmaeilzadeh et al. 2013). So, the effect of miniaturization has driven us to the fabrication of billions of transistors integrated on a single chip, which are not all used all the time, since they could be destroyed due to thermal dissipation. Another problem is the ways in which algorithms are used and implemented. In computers, we are using generally sequential algorithms, and sequential computing. A simple example is provided by the problem of finding the maximum among eight numbers from 1 to 8 written in an arbitrary order. To solve this problem, a computer needs seven steps/comparing operations for searching progressively the biggest number between adjacent numbers. An example is given in Fig. 4.2.

On the other hand, parallel computation needs less time to compute. In the example above, the first step would be to find directly the maximum among the four numbers in the first sequential step, followed by other two similar steps needed to solve the problem: in total one needs 3 steps. The time necessary in parallel computation is on the order of $\log_2 N$, i.e., $\log_2 8 = 3$ steps, N being in this case the number of entries in the list, while for sequential computation the computing time is proportional to N. Note, however, that there are classical computers that are able to perform parallel computations, an example in this respect being computers with multiple processors (which need to be perfectly synchronized), each of them running a different program on its own set of data.

Parallel computation can be directly associated to the evolution of a quantum system described by a wavefunction representing a superposition of states. This is

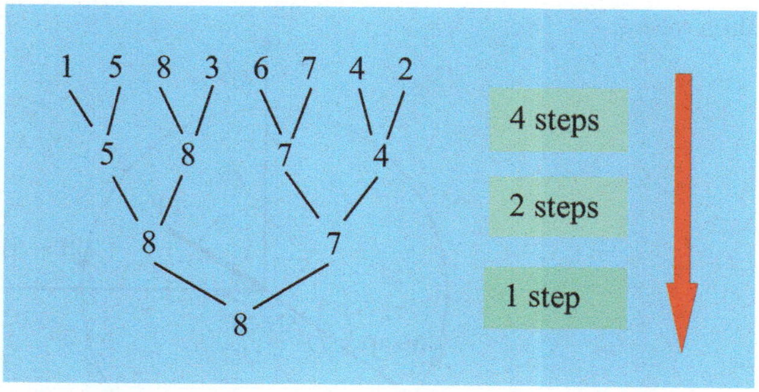

Fig. 4.2 Example of sequential computing: finding the maximum among eight numbers written in an arbitrary order

why, despite tremendous technological challenges, the field of quantum computing has become one of the major areas of interest in physics and technology. There are many excellent monographies on the principles of quantum computing (see, for instance Nielsen and Chuang 2010; Stolze and Suter 2004), which can be consulted by readers interested in understanding the theory of quantum computation and the associated conceptual and technological challenges. Therefore, we will not enter into details regarding this subject, but only present the essential facts without which the understanding of quantum logic devices would not be possible.

Quantum bits, or qubits, are the quantum counterparts of the 0 and 1 classical bits, and represent distinguishable different states of a quantum system with a certain complex wavefunction $|\Psi\rangle$. The quantum wavefunction can generally be in a superposition of these $|0\rangle$ and $|1\rangle$ states: $|\Psi\rangle = a|0\rangle + b|1\rangle$, and not only in one or the other of these states, in contrast to a classical state, which can be in only one state: the bit 0 or 1. Thus, there is a finite probability to find the quantum system in each of the two basis states $|0\rangle$ and $|1\rangle$: there is a probability of $|\langle\Psi|0\rangle|^2 = |a|^2$ to find the quantum system in the $|0\rangle$ state, and a probability $|\langle\Psi|1\rangle|^2 = |b|^2$ to find the quantum system in the $|1\rangle$ state. We have always $|a|^2 + |b|^2 = 1$.

If the wavefunction refers to a two-qubit state, then the wavefunction is a superposition of four states: $|\Psi\rangle = a|00\rangle + b|01\rangle + c|10\rangle + d|01\rangle$, with a, b, c, d complex numbers and $|xy\rangle = |x\rangle \otimes |y\rangle$, with $x, y = 0, 1$ and $\otimes$ indicating the tensor product of qubit states, is the state in which the first qubit has the logical value x and the second the logical value y. An n-qubit state is a superposition of 2^n states.

The quantum state is a vector. Since the state $|\Psi\rangle$ is not changed under a phase shift, i.e., $\exp(i\phi)|\Psi\rangle$ is equivalent to/represent the same state $|\Psi\rangle$ for any real number ϕ, it follows that we can always chose as a a real number such that, via the relation between the moduli of a and b, we can relate a and b via a single parameter. More precisely, we can express the quantum wavefunction as

$$|\Psi\rangle = \cos(\theta/2)|0\rangle + \exp(i\phi)\sin(\theta/2)|1\rangle \tag{4.1}$$

Fig. 4.3 The Bloch sphere

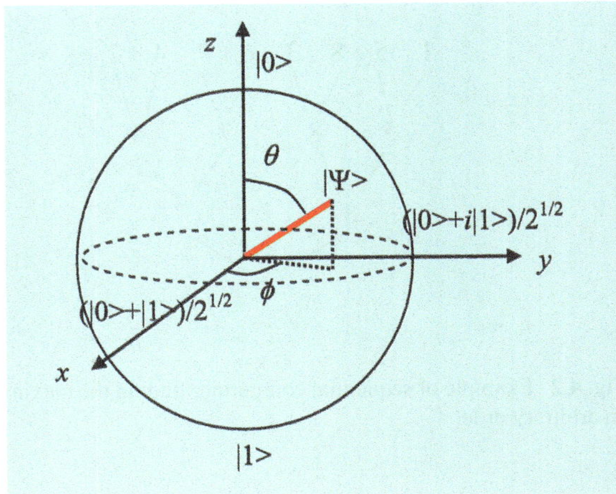

relation that indicates the possibility to establish a bijective relation between the qubits described by (4.1) and the states on a sphere, called Bloch sphere, where the θ and ϕ represent the spherical coordinates (see Fig. 4.3).

The state $|\Psi\rangle$ of a single qubit can then be regarded as a point on the surface of the sphere, having unitary radius 1, point uniquely described by the two angles θ and ϕ. In particular, $\theta = 0$, associated to the north pole of the Bloch sphere in Fig. 4.3, is thus identified with $|\Psi\rangle = |0\rangle$, case in which the south pole, for which $\theta = \pi$, corresponds to $|\Psi\rangle = |1\rangle$.

In many situations it is more convenient to represent the state of a quantum system as a vector $\begin{pmatrix} a \\ b \end{pmatrix}$ defined in the complex space C^2 spanned by $|0\rangle$ and $|1\rangle$, or, equivalently, as $(a, b)^T$, where ^{T} denotes the transposition operation. The states $|0\rangle$ and $|1\rangle$ are orthogonal and form the basis state for computation. The basis state is not uniquely defined. For example, if $|\Psi\rangle$ is a qubit state and $|\Phi\rangle$ is another state perpendicular to it, so that $\langle\Phi|\Psi\rangle = 0$, there is a unique unitary transformation which transform the states $|0\rangle$ and $|1\rangle$ into $|\Psi\rangle$ and $|\Phi\rangle$, which can be considered as another basis state for computation.

In addition of being associated to quantum states, which tend to lose their coherence rapidly in time, the qubits must interact between them in order to implement logical operations, and must be accessible for manipulation and reading of their initial, final and control states. These requirements pose both theoretical and practical challenges but, at the same time, open up new avenues of investigation.

The quantum systems are inherently parallel since a quantum state subjected to an interaction, which must be designed as analogous to a computing algorithm, is describing simultaneously the evolution of all possible states. The major problem of quantum computation is the rapid loss of coherence of quantum states, termed as decoherence. If a quantum system is losing its coherence, the relevancy and accuracy of computation is lost. Quantum coherence is lost by interaction with the

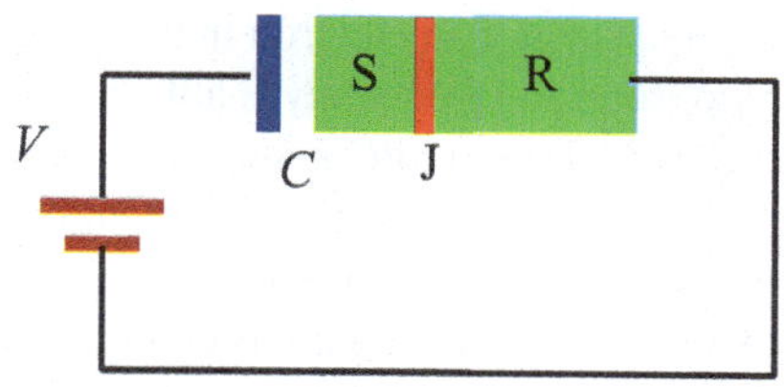

Fig. 4.4 The Josephson qubit

environment. Therefore, the qubits are generated in quantum systems with reduced number of states/degrees of freedom and with discrete energy levels that should be well separated; in this way, decoherence is minimized. Some examples of qubits implementation, among many existing in the literature, are presented briefly in the following.

The first example is that of a Josephson junction, which is the key element of all quantum computers working today at milliKelvin temperatures (see Fig. 4.4). Josephson qubits are implemented using a reservoir R, which is a superconductor, separated from another superconductor region S by and oxide/Josephson junction J. The region S placed between the gate capacitor and the junction J plays the role of an island. By changing the bias *V*, the electrostatic energy of the island S is changing, and the transfer of a single Cooper pair of electrons from the reservoir to the island or from the island to the reservoir by tunneling though the junction J. For a certain value of the bias *V*, only two states/Cooper pair, which are in excess in the island and are characterized by quantum states with number of electrons/electric charges |*n*· and |*n* + 1·, are contributing to the Cooper pair dynamics. The quantum states of qubits are thus the superposition of charges of these two states. In this case, the detection of qubits can be done with single-electron transistors (SETs), which can detect very small changes of charges by observing the oscillations in time of Cooper pairs. Few oscillations are occurring, however, due to decoherence, unless the device is working at milliKelvin temperatures.

Qubits can be implemented also in a quantum dot, as that shown in Fig. 4.5(a). In this example, the qubits are encoded in the fundamental state and in the first

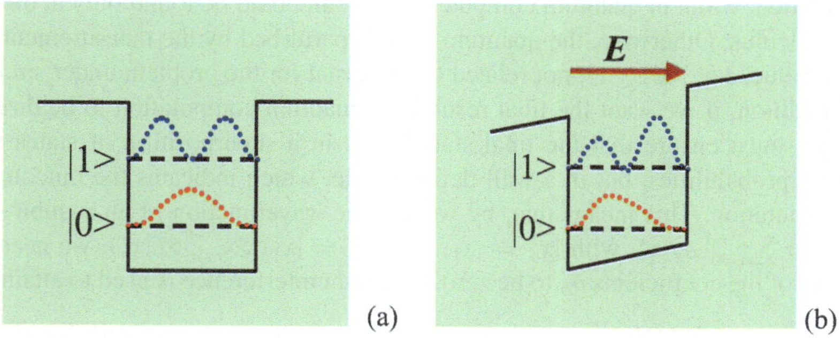

Fig. 4.5 (a) Qubits in a quantum dot, and (b) state transformation in the presence of an electric field, necessary for implementing the CNOT transformation

excited state of an electron in the quantum dot. An electromagnetic field at a suitable amplitude and frequency is able to change the electron state from the fundamental to the excited state or in a superposition of these two states, similarly to Rabi oscillations. To implement a CNOT transformation, a static field E (produced by a gate) is applied to a neighboring quantum dot [see Fig. 4.5(b)], such that the distribution of charge in the initial quantum dot is changed due to the dipolar interaction between quantum dots; the neighboring quantum dot is now in an excited state. As a result, the energy levels of the fundamental and excited states of the initial quantum dot, |0· and |1·, as well as the frequency of the electromagnetic field that can induce transitions between them, are modified. Thus, the neighboring quantum dot controls the state/the transitions between the energy levels of the initial quantum dot taking place at a given frequency. More precisely, if f_0 is the transition frequency in the absence of the static electric field on the neighboring quantum dot and f_1 the transition frequency between states in the initial quantum dot when the electric field is removed, if we apply an electromagnetic pulse with frequency f_1 on the initial quantum dot, the electron performs a transition only if the neighboring quantum dot is in an excited state/a static electric field is applied on it. This interaction embodies the quantum CNOT operation, or logic gate.

4.1.2 The Processing and Measurement of Qubits

Quantum computers work using reversible operations U, which transform the initial state of the qubits towards a final state. These operations are linear and unitary, transforming a state with modulus 1 in Hilbert space into another state having the same property. The single irreversible process in quantum computing is the measurement, which is the only possibility to extract the desired information after the qubits take the final value, which represents the outcome of the computation. To read the outcome, an additional qubit, which takes a defined logical value at the end of the computation, is used in order to be sure that the quantum computation is ended and the result can be read. In classical computers, we can read at any time the intermediate results of a computation, while in quantum computers the result could be found only at the end of the calculus. Otherwise, the quantum state is perturbed by the measurement and takes a value, e.g. $|00\ldots 1\rangle$, not related to/irrelevant for the problem under study.

In addition, if we want the final result of a quantum computation to be directly read, we must ensure that the final state is not in a superposition of states with different probabilities, but in a well-defined state, which indicates the outcome of the computation. This means that, by writing the wavefunction of an n-qubit state as $|\Psi\rangle = \sum_{x=0}^{2^n-1} a_x|x\rangle$, with $|x\rangle = |x_1, \ldots, x_n\rangle = |x_1\rangle \otimes \ldots \otimes |x_n\rangle$, we need the majority of the coefficients a_x to be zero. Quantum interference is used to attain this goal.

Another striking difference between bits and qubits is the fact that arbitrary qubits cannot be copied. This fact complicates the quantum computation, but represents the major advantage of quantum information transfer/communication. There is a

Table 4.1 Main differences between classical and quantum computation

Classical computation	Quantum computation
Bit: 0 or 1	Qubit: superposition of quantum states $a\|0\rangle + b\|1\rangle$
A n-bit register represents a single information bit	A n-qubit register represents a superposition of 2^n states
The state of classical bits can be verified at any moment in the memory	We can measure the quantum state only at the end of the computation
There is no correlation (entanglement) between bits	There can be entanglement between qubits at any distance; separate parts of a quantum computer could be linked together, and teleportation is possible
The bits can be copied	Arbitrary qubits cannot be copied
Logical gates: generally irreversible and based on Boolean algebra	Logical gates: unitary and reversible operators
Universal gates (gates that can be used to implement any logical operation): NAND or NOR	Universal gates: CNOT and 1 qubit gates
The reading of the final computation result is made in a single step	Quantum interference is used to read the final result in an efficient way

theorem, the so-called "no-cloning" theorem, which demonstrates easily this fact (Nielsen and Chuang 2010). The non-cloning theorem shows once again the weirdness of the quantum world compared to the classical world. Although information is copied in a huge number of versions in classical computers and biological systems, this basic operation is forbidden in the quantum systems. The main differences between classical and quantum computation are summarized in Table 4.1.

4.1.3 Logical Quantum Gates

The only reversible operation in classical computers acting on a single bit is NOT. On the contrary, in the case of a quantum computer the reversible operations on a single qubit are all linear and unitary transformations, which can transform a unity-modulus vector into another. The NOT operation is described by the following unitary matrix:

$$U_{\text{NOT}} = \begin{pmatrix} 0 & 1 \\ 1 & 0 \end{pmatrix}. \tag{4.2}$$

which can be obtained by representing the qubits $|0\rangle$ and $|1\rangle$ as $\begin{pmatrix} 1 \\ 0 \end{pmatrix}$ and $\begin{pmatrix} 0 \\ 1 \end{pmatrix}$, respectively and making use of the known result of the operation: $U_{\text{NOT}}|0\rangle = |1\rangle$ and $U_{\text{NOT}}|1\rangle = |0\rangle$.

A more general one-qubit operation without classical analogue is $U_A = 2^{-1/2}\exp(i\pi/4)(I_2 - i\sigma_x)$, where I_2 denotes the unitary matrix and σ_x is one of the Pauli spin matrices, introduced in quantum mechanics to describe the angular momentum of an electron with spin 1/2:

$$\sigma_x = \begin{pmatrix} 0 & 1 \\ 1 & 0 \end{pmatrix}, \quad \sigma_y = \begin{pmatrix} 0 & -i \\ i & 0 \end{pmatrix}, \quad \sigma_z = \begin{pmatrix} 1 & 0 \\ 0 & -1 \end{pmatrix}. \tag{4.3}$$

From (4.2) and (4.3) it follows that $U_{\mathrm{NOT}} = \sigma_x$ and that if we apply U_A two times the result is the NOT operation, so that this operator is named $U_A = U_{(\mathrm{NOT})^{1/2}}$ or as $U_{\sqrt{\mathrm{NOT}}}$. Note that the unity operator I_2 and the three Pauli matrix operators form a basis for the bidimensional matrix algebra, in the sense that any operator can be written as $U = U_0 I_2 + U_x\sigma_x + U_y\sigma_y + U_z\sigma_z$.

Another quantum gate without classical analogue acting on one-qubit states is the Hadamard gate, which transform the state $|0\rangle$ in the superposition $2^{-1/2}(|0\rangle + |1\rangle)$ and the state $|1\rangle$ in $2^{-1/2}(|0\rangle - |1\rangle)$. In a similar way as above, the action of this gate is described by $U_{\mathrm{H}} = 2^{-1/2}(\sigma_x + \sigma_z)$ or, in matrix form.

$$U_H = \frac{1}{\sqrt{2}} \begin{pmatrix} 1 & 1 \\ 1 & -1 \end{pmatrix} \tag{4.4}$$

If we apply twice the Hadamard we get the identity operator, so that $U_H = U_{(I_2)^{1/2}}$. The symbols for the above mentioned gates are represented in Fig. 4.6.

Note that by applying n elementary operators such as Hadamard gates on a n-qubit register, we can obtain a state with all 2^n possible numerical values of the register, while if we apply n elementary operators to a classical register we prepare only a single state of the register, which represents a single number. So, the quantum gates are used not only for quantum computing, but also for the preparation of a desired quantum state, in particular for the preparation of the initial state of the quantum system entering the computing process. For example, if we apply to every qubit in the state $|0\rangle|0\rangle = |00\rangle$ the Hadamard transform, we obtain:

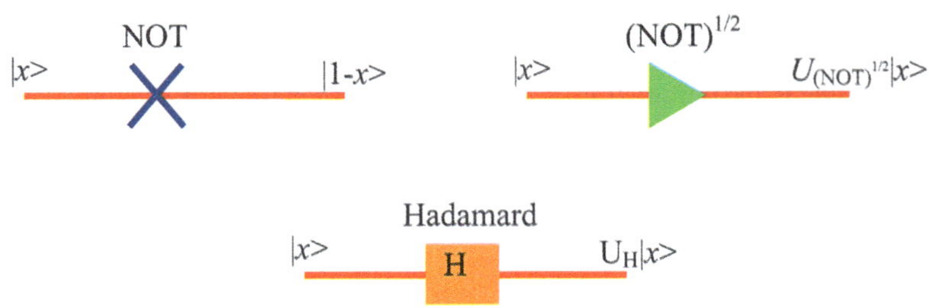

Fig. 4.6 Graphical representations of the NOT, $(\mathrm{NOT})^{1/2}$, and Hadamard gates

$$(H \otimes H)(|0\rangle|0\rangle) = (H|0\rangle)(H|0\rangle) = \frac{1}{\sqrt{2}}(|0\rangle + |1\rangle)\frac{1}{\sqrt{2}}(|0\rangle + |1\rangle)$$
$$= \frac{1}{2}(|00\rangle + |01\rangle + |10\rangle + |11\rangle) \quad (4.5)$$

If we generalize this operation to the n-times tensorial product of Hadamard transforms, the operator $H^{\otimes n}$, applied to the state $|0\rangle_n$ of n qubits, the result is the superposition with equal weights of all possible states of n qubits:

$$H^{\otimes n}|0\rangle_n = \frac{1}{2^{n/2}} \sum_{0 \le x < 2^n} |x\rangle_n \quad (4.6)$$

In classical computers, the most general reversible operation is the permutation of 2^n distinct states, and the result is $(2^n)!$ possible distinct operations. In a quantum computer, the most general reversible operation on n-qubits state is a 2^n-dimensional unitary transformation. An operator is acting in the Hilbert space C^{2^n} and can be represented by a unitary $2^n \times 2^n$ matrix. These transformations are n-qubit gates.

One of the most used gates in quantum computing is CNOT (control-NOT), which changes the state of a qubit (called signal qubit or target qubit) only if another qubit, termed as control qubit, is in the state $|1\rangle$. This gate is acting on 2-qubit states as follows

$$U_{\text{CNOT}}|00\rangle = |00\rangle, \quad U_{\text{CNOT}}|01\rangle = |01\rangle, \quad U_{\text{CNOT}}|10\rangle = |11\rangle, \quad U_{\text{CNOT}}|11\rangle = |10\rangle \quad (4.7)$$

such that its matrix representation is

$$U_{\text{CNOT}} = U_{\text{XOR}} = \begin{pmatrix} 1\,0\,0\,0 \\ 0\,1\,0\,0 \\ 0\,0\,0\,1 \\ 0\,0\,1\,0 \end{pmatrix} \quad (4.8)$$

The matrix in (4.8) corresponds to the case when the control qubit is the first qubit (it interchanges the states $|10\rangle$ and $|11\rangle$). If the control qubit is the second one, so it interchanges the states $|01\rangle$ and $|11\rangle$, the matrix form of CNOT becomes

$$U_{\text{CNOT}} = \begin{pmatrix} 1\,0\,0\,0 \\ 0\,0\,0\,1 \\ 0\,0\,1\,0 \\ 0\,1\,0\,0 \end{pmatrix} \quad (4.9)$$

The CNOT gate can be expressed also in terms of the Pauli matrices as:

$$U_{\mathrm{CNOT}} = |0\rangle\langle 0| \otimes I_2 + |1\rangle\langle 1| \otimes U_{\mathrm{NOT}} = 2^{-1}(1+\sigma_z) \otimes I_2 + 2^{-1}(1-\sigma_z) \otimes \sigma_x \tag{4.10}$$

Other gates that act on 2-qubit states and have no classical analogue are Cph (control phase), which imparts a phase of ϕ on the target qubit when both control and target qubits have logical states of 1, SWAP, and $(\mathrm{SWAP})^{1/2}$. The SWAP gate exchanges the states of two qubits. The matrices of these 2-qubit gates are, respectively,

$$U_{\mathrm{Cph}}(\phi) = \begin{pmatrix} 1\,0\,0 & 0 \\ 0\,1\,0 & 0 \\ 0\,0\,1 & 0 \\ 0\,0\,0 & \exp(i\phi) \end{pmatrix}, \quad U_{\mathrm{SWAP}} = \begin{pmatrix} 1\,0\,0\,0 \\ 0\,0\,1\,0 \\ 0\,1\,0\,0 \\ 0\,0\,0\,1 \end{pmatrix},$$

$$U_{\sqrt{\mathrm{SWAP}}} = \begin{pmatrix} 1 & 0 & 0 & 0 \\ 0 & (1+i)/2 & (1-i)/2 & 0 \\ 0 & (1-i)/2 & (1+i)/2 & 0 \\ 0 & 0 & 0 & 1 \end{pmatrix} \tag{4.11}$$

The above-mentioned 2-qubit gates have standard graphical representations shown in Fig. 4.7. The black point in the CNOT and the Cph gates means that the gate is acting only if the control qubit has the logical value 1. If the circle is empty it means that the action of the gate occurs for the 0 value of the control qubit.

There are other gates which are acting on 3-qubits states, such as the Toffoli gate, termed as CCNOT or C^2NOT, the Fredkin gate or CSWAP (controlled SWAP), and the Fredkin-Toffoli gate, which is another type of CSWAP. The Toffoli gate changes the vale of the signal bit (from 0 to 1 or from 1 to 0), if the other two bits are 1, while the Fredkin-Toffoli gate exchanges the logical values of bit2 and bit3 when bit1 is one (see below the truth Table 4.2 for the Fredkin-Toffoli gate). The Fredkin gate

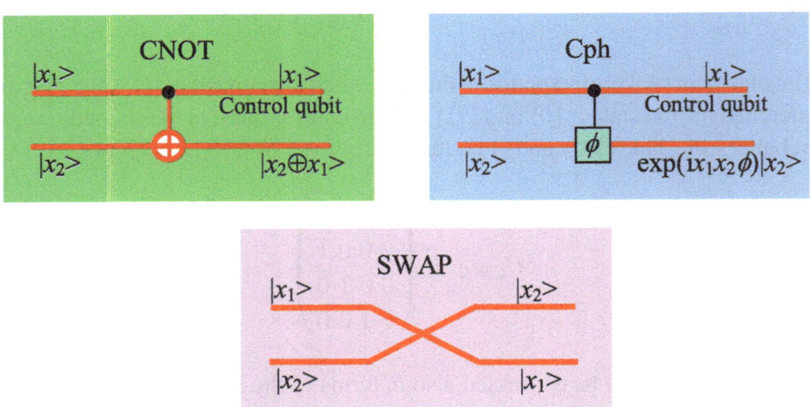

Fig. 4.7 Representations of the CNOT, Cph, and SWAP gates

Table 4.2 The truth table of Fredkin-Toffoli gate

bit1 bit2 bit3	Fredkin-Toffoli
0 0 0	0 0 0
0 0 1	0 0 1
0 1 0	0 1 0
0 1 1	0 1 1
1 0 0	1 0 0
1 0 1	1 1 0
1 1 0	1 0 1
1 1 1	1 1 1

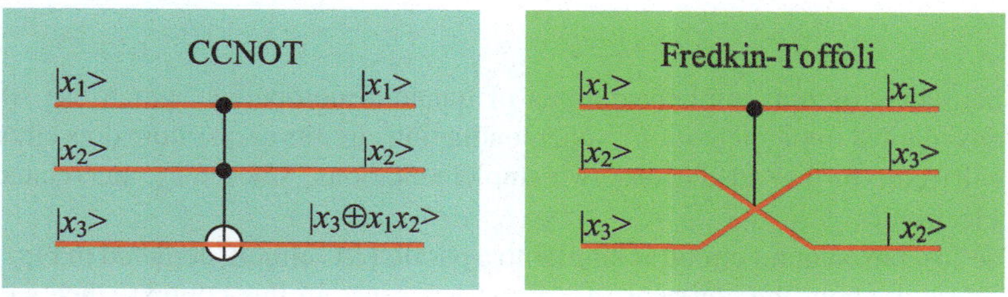

Fig. 4.8 The graphical representation of CCNOT gate and Fredkin-Toffoli

has a similar truth table as the Fredkin-Toffoli gate, exchanging between them the logical value of two input qubits if the value of the control qubit is zero. The graphical representations of the CCNOT (Toffoli) and Fredkin-Toffoli gates are illustrated in Fig. 4.8.

It can be demonstrated that any quantum logical gate can be decomposed in a succession of elementary quantum gates: an arbitrary rotation of 1-qubit state and a 2-qubit CNOT gate. As such, the rotation gate and CNOT form a set of universal gates. For example, an SWAP gate S_{ij}, where i is the control qubit and j is the signal qubit, can be decomposed in CNOT operations $C_{ij} : S_{ij} = C_{ii}C_{ji}C_{ij}$, while the operation that exchanges the control and signal qubits in CNOT, i.e. transforms C_{ij} into C_{ji}, can be decomposed in turn in a succession of Hadamard gates H_i applied to the qubit i and CNOT gates: $C_{ji} = (H_i H_j)C_{ij}(H_i H_j)$.

Quantum computation is working much faster than classical calculus for a series of algorithms, such as Deutsch-Jozsa, Grover, or quantum Fourier transform, which are very well-described in Nielsen and Chuang (2010) and thus will not be repeated here. In some algorithms, entanglement or state correlation is used, a physical phenomenon which is specific only to quantum systems, and is described in standard books on quantum computation. It is not sure, however, that every program executed by a classical computer will work much faster on a quantum computer; in fact, there is no method to transform a classical algorithm that solves a certain problem into a quantum counterpart.

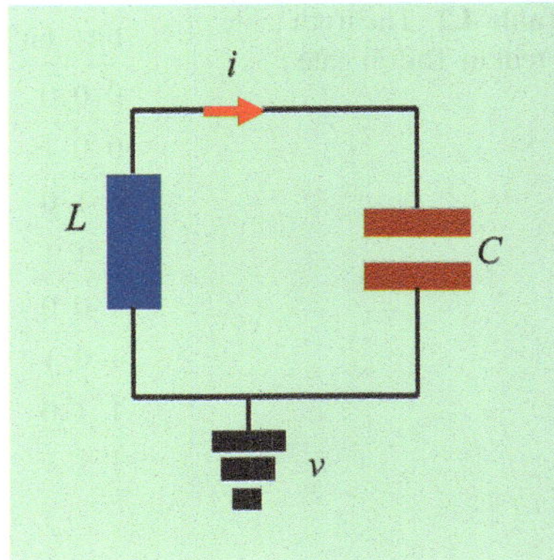

Fig. 4.9 Classical LC circuit

There are different implementations of quantum computers, which use single photons (optical implementation), superconducting circuits or quantum dots circuits. We will refer further to the last two implementations, since they are related to nanoelectronics.

The Josephson qubit circuit is similar to a parallel LC circuit, depicted in Fig. 4.9. A discussion about the classical LC circuit has sense in this context since such a circuit can be described at the classical level with a Hamiltonian analogous to a mechanical oscillator of mass m and resonant frequency ω, the latter having the form $H = p^2/2m + m\omega^2x^2/2$. More precisely, the Hamiltonian of the parallel LC circuit is given by

$$H = (1/2)CV^2 + (1/2)LI^2 \tag{4.12}$$

If the mechanical conjugate variables momentum, p, and displacement, x, are analogous to the electrical parameters voltage, V, and current, I, the two Hamiltonians can be put into correspondence if the mass m is replaced by the capacitance C and the resonant frequency by $1/\sqrt{LC}$ (Krantz et al. 2019).

Further, the quantum description of the parallel LC circuit involves two new coordinates: the reduced flux, $\phi = 2\pi\Phi/\Phi_0$, with $\Phi(t) = \int_{-\infty}^{t} V(t')\mathrm{d}t'$ and $\Phi_0 = h/(2e)$, and the reduced charge, $n = Q/2e$, where Q is the charge of the capacitor. Note that Q is the momentum conjugate to the flux in the Lagrangian sense, since $Q = C\dot{\Phi}$. The quantum Hamiltonian of the parallel LC circuit is given by

$$H = 4E_C n^2 + (1/2)E_L\phi^2 \tag{4.13}$$

where $E_C = e^2/2C$ is the energy needed to add each electron of the Cooper pair and $E_L = (\Phi_0/2\pi)^2/L$ represents the inductive energy. The two operators, n and ϕ, satisfy the commutation relation $[\phi, n] = i$. The first term in the quantized Hamiltonian (4.13) can be interpreted as kinetic energy, while the second can be seen as potential energy.

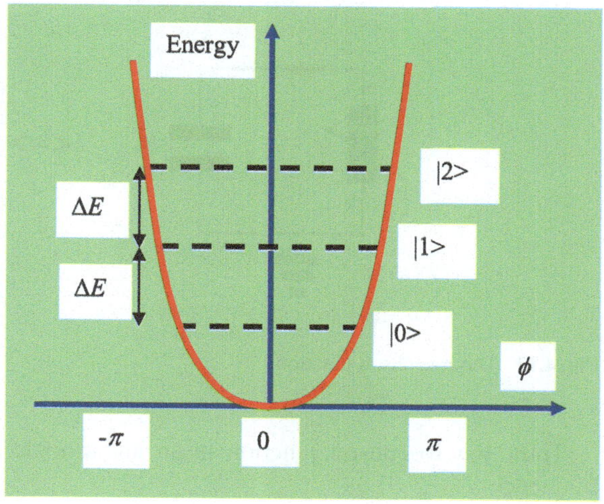

Fig. 4.10 The energy potential of the quantized LC parallel circuit

The quantized Hamiltonian of the parallel LC circuit describes one particle in a quadratic potential, i.e., it is of a quantum harmonic oscillator-type, and thus has discrete energy levels, corresponding to the distinguishable levels of qubits:

$$\Delta E = E_{n+1} - E_n = \hbar\omega_r \tag{4.14}$$

where $\omega_r = 1/(LC)^{1/2} = (8E_L E_C)^{1/2}/\hbar$.

The dependence of the potential energy on ϕ is shown in Fig. 4.10.

If we replace the inductance of the parallel LC circuit with a Josephson junction, we replace a linear inductance with a nonlinear inductance, represented by the superconducting junction. As mentioned previously, the Josephson junction is formed from a very thin dielectric, of few nanometers thickness (for example, Al_2O_3), sandwiched between two superconductors, such as Al. The Cooper pairs are tunneling the dielectric, so that the Josephson junction is a superconducting tunnel diode.

The Josephson current and voltage relations are given by

$$I = I_C \sin(\phi), \quad V = (\hbar/2e)(\mathrm{d}\phi/\mathrm{d}t) \tag{4.15}$$

where I_C is the maximum superconducting current, the phase ϕ being defined in the presence of a voltage V as $\hbar\mathrm{d}\phi/\mathrm{d}t = 2eV$.

Performing a time derivative of the current in (4.15), one observes that the Josephson junction is a nonlinear inductor, with inductance (Clarke and Wilhelm 2008)

$$L_i = \Phi_0/(2\pi I_C \cos\phi) = \Phi_0/\sqrt{2\pi I_C^2 - I^2} \tag{4.16}$$

A Josephson junction has always a capacitance in parallel with the inductance, and is represented as shown in Fig. 4.11.

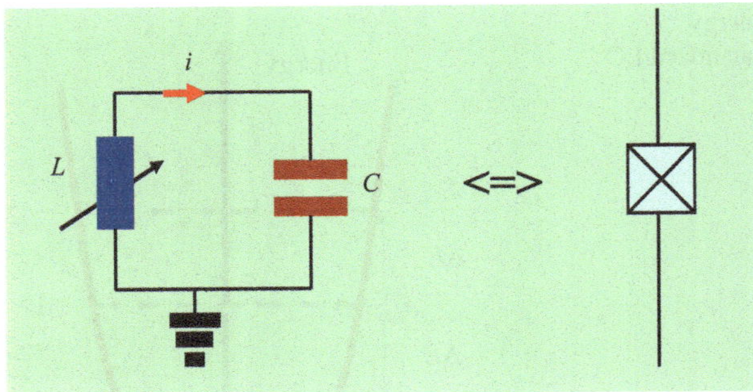

Fig. 4.11 The Josephson junction

Thus, the Josephson junction is an anharmonic oscillator, with the resonant frequency

$$\omega = 1/\sqrt{L_i C} = \sqrt{2\pi I_C \Phi_0 C}/(1 - I^2/I_C^2)^{1/4}, \tag{4.17}$$

this nonlinear oscillator being described by the Hamiltonian (Krantz et al. 2019)

$$H = 4E_C n^2 - E_J \cos(\phi), \tag{4.18}$$

where

$$E_C = e^2/(2C_T) \tag{4.19}$$

with C_T the total capacitance formed from the capacitance C of the junction and a shunt capacitance, the role of which will become clear below, and

$$E_J = I_C \Phi_0/2\pi \tag{4.20}$$

the Josephson energy. In this anharmonic oscillator, the representation of the energy dependence on the superconducting phase ϕ shows still discrete levels, which are, however, unequally distanced, since its potential energy is no longer parabolic, but of a sinusoidal shape.

This anharmonic oscillator can be represented with the help of the Hamiltonian

$$H = \omega_q \sigma_z/2 \tag{4.21}$$

where σ_z is the Pauli z operator and ω_q are the resonant frequencies described above and expressed as $\omega_q = (\sqrt{8E_C E_J} - E_C)/\hbar$.

From the energy dependence on phase, represented in Fig. 4.12, it follows that one can use the lowest energy levels as the quantum states $|0\rangle$ and $|1\rangle$ of a qubit. The Josepshon qubits are excited by a frequency ω_q, where q is 0 or 1, situated in the

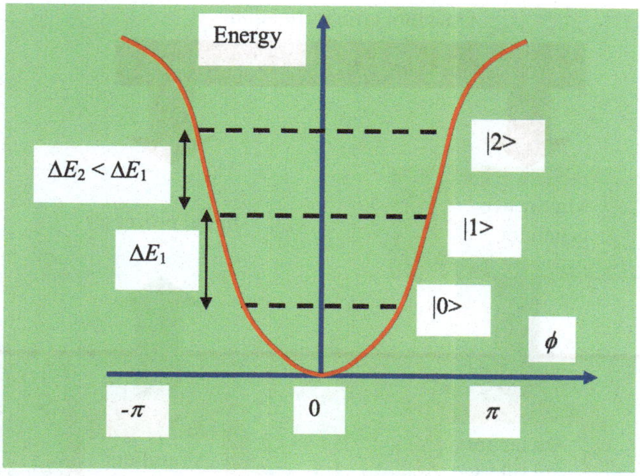

Fig. 4.12 Energy dependence on phase in Josephson junction

microwave range of 3–8 GHz, which can be tuned by a DC voltage and very precise microwave generators.

The condition that the ratio $E_J/E_C \geq 50$ is necessary to decrease and disable any charge influence, i.e., any noise, and thus to increase the coherence of the quantum oscillator. This operation is performed with the help of a Josephson junction in parallel with a shunted capacitance, or with two parallel Josephson junctions.

Based on the considerations above, one can imagine now the structure of a quantum computer based on Josephson qubits; a schematic representation of it is shown in Fig. 4.13 (Gambetta et al. 2017). The Josephson junctions are fabricated using e-beam lithography and need milliKelvin temperatures and low noise environment to work properly. Thus, a quantum computer based on Josephson qubits is a superconducting microwave circuit working at very low temperatures. The Google, IBM, Amazon and other quantum computers work in this configuration. The Josephson qubits are programmable by various microwave frequencies in the range of 4–7 GHz associated with tuning the Josephson energy via magnetic fields generated by external currents.

There are two temporal parameters that characterize the performance of quantum computers: the relaxation time T_1, which is the time required for a qubit to relax from the (first) exited state $|1\rangle$ to the ground state $|0\rangle$, and the dephasing time T_2, which is the time needed for the superposition state $(|0\rangle + |1\rangle)/\sqrt{2}$ to lose the phase information; both times should be as long as possible. In the modern IBM 20-qubit quantum computers these times are about 7–100 μs.

An example of a qubit based on Josephson junctions involves transmons. The transmon is a Josephson circuit containing two Josephson junctions in parallel, as shown in Fig. 4.14, in order to render the circuit insensitive to the 1/*f* noise (Houck et al. 2009). This is achieved by implementing the condition $E_J \gg E_C$, i.e., by decreasing the charging energy $E_C = e^2/(2C_T)$ via increasing the total capacitance.

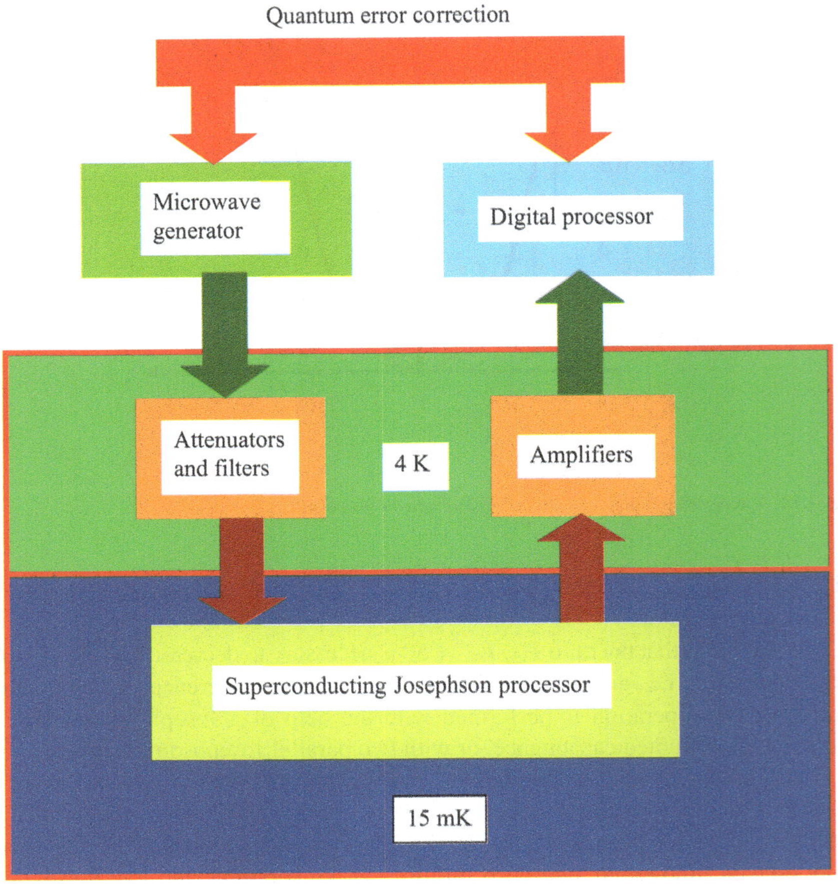

Fig. 4.13 The superconducting Josephson quantum computer

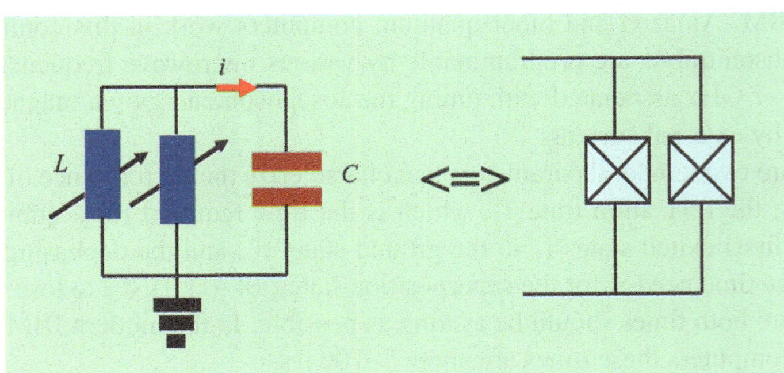

Fig. 4.14 The transmon

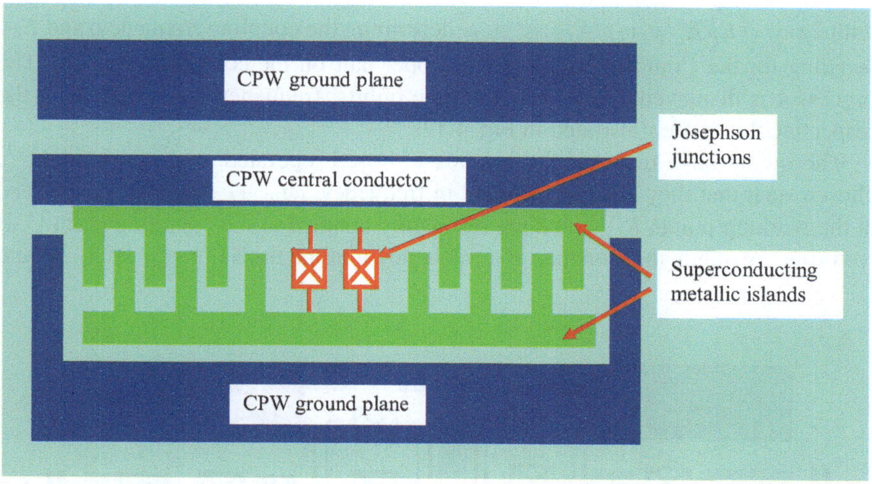

Fig. 4.15 The transmon in cavity CPW technique

The typical transmon circuit structure is based on coplanar waveguide (CPW) and advanced microwave circuits, as illustrated in Fig. 4.15. The interdigitated metallic structure and the relative large dimensions of the superconductive islands increase very much the inter-island capacitance; thus, the transmon is lowering E_C via the large shunt capacitance. The CPW consists of three metallic parallel strips separated by two equal gaps. The outer metallic strips are grounded, while the central conductor is the signal electrode. The impedance of the CPW is 50 Ω and the gaps are 3–4 μm wide. The outer electrodes are made of Nb, while the interdigitated capacitance is made of Al, the same metal used for the Josephson junction ($Al/Al_2O_3/Al$).

We present now a circuit consisting of two transmons that form a processor. Each transmon $j = $ I,II, is a capacitively shunted SQUID (Dewes et al. 2012), having the total Josephson energy

$$E_J^j(\phi_j) = E_J^j|\cos(x_j)|\sqrt{1 + d_j^2 \tan^2(x_j)} \tag{4.22}$$

where d_j accounts for the asymmetry between the Josephson junctions and $x_j = \pi\Phi_j/\Phi_0$, Φ_j being the magnetic flux through the SQUID induced by two currents with 0.5 GHz bandwidth. The transmon frequencies, denoted as ν_j are dependent on $E_J^j(\phi_j)$, and are thus tunable via Φ_j, by the current lines I_j.

$$\nu_j \to |0\rangle, |1\rangle \propto I_j \tag{4.23}$$

The qubits are capacitively coupled and are described by the Hamiltonian

$$H = h(-\nu_I\sigma_z^I - \nu_{II}\sigma_z^{II} + 2g\sigma_y^I\sigma_y^{II})/2 \tag{4.24}$$

with $2g = \sqrt{E_C^I E_C^{II} \nu_I \nu_{II} / E_{Cc}} \ll \nu_{I,II}$ describing the coupling frequency, and E_{Cc} standing for the Coulomb energy of a Cooper pair on the coupling capacitor. The processor is displayed in Fig. 4.16, and the qubits' frequency dependences on the respective fluxes are illustrated in Fig. 4.17.

The qubits have resonant frequencies $\nu_R^{I,II}$, are operated at the points $M^{I,II}$, chosen such that they are well detuned in frequency, i.e., $\nu_{II} - \nu_I \gg 2g$, are read at the working points $R^{I,II}$, and are resonantly coupled at the operating point C. If we bring the qubits in resonance at a frequency ν in the time $1/\nu \ll t \ll 1/2\,g$ and

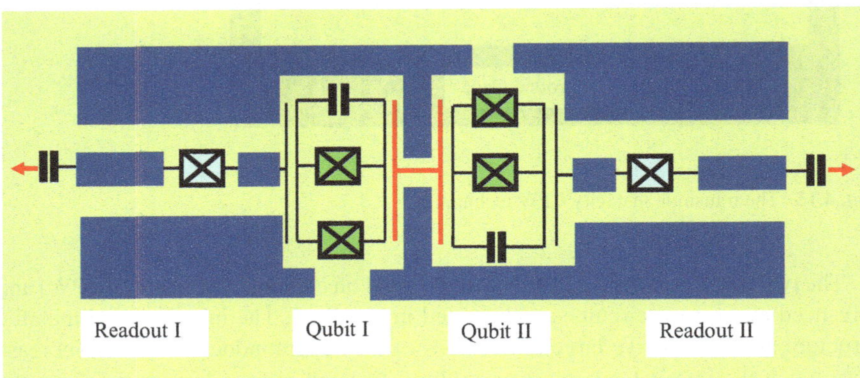

Fig. 4.16 Josephson processor with two transmons as qubits

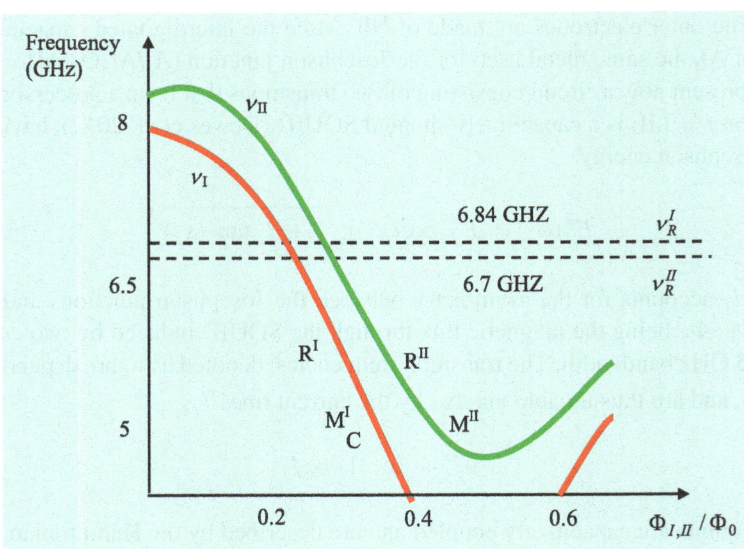

Fig. 4.17 The qubits' frequency dependence on the induced flux

keep them in this state for a time duration Δt, we can implement the operation

$$O = \Theta_I \Theta_{II} (\sqrt{iSWAP})^{8g\Delta t} \tag{4.25}$$

where $\Theta_j = \exp(i\theta_j \sigma_z^j/2)$, with $\theta_j = 2\pi(\nu - \nu_j)dt$ the dynamical phase which occurs during coupling.

Thus, the processor is an exact $\sqrt{iSWAP}$ gate when $\Delta t = 1/8g$ and after a rotation Θ_j^{-1}. In these conditions, we have

$$U_{\sqrt{\text{iSWAP}}} = \begin{pmatrix} 1 & 0 & 0 & 0 \\ 0 & 1/\sqrt{2} & -i/\sqrt{2} & 0 \\ 0 & -i/\sqrt{2} & 1/\sqrt{2} & 0 \\ 0 & 0 & 0 & 1 \end{pmatrix} \tag{4.26}$$

For the readout, each qubit is capacitively coupled to a $\lambda/2$ coplanar wave resonator with a quality factor of $Q = 700$, and each resonator is made nonlinear using a separate Josephson junction. Microwave homodyne detection is used and the outcome is a high-fidelity universal gate $\sqrt{iSWAP}$.

Two-qubit quantum algorithms such as Grover or Deutsch-Jozsa and Grover can be experimentally demonstrated with a four-port processor containing two transmons and a flux-bias line located near the qubits (DiCarlo et al. 2009). Denoting by $|l, r\rangle$ the quantum state defined by the excitation levels l and r of the two (left and right) transmon qubits, their interaction takes place when the $|1, 1\rangle$ and $|0, 2\rangle$ states become degenerate, the associated frequency shift needed to avoid crossing producing the phase gate given by

$$U = \begin{pmatrix} 1 & 0 & 0 & 0 \\ 0 & \exp(i\phi_{01}) & 0 & 0 \\ 0 & 0 & \exp(i\phi_{10}) & 0 \\ 0 & 0 & 0 & \exp(i\phi_{11}) \end{pmatrix} \tag{4.27}$$

where ϕ_{lr} is the dynamical phase acquired by $|l, r\rangle$. The controlled-phase gate U_{Cph} is the implemented with the help of a pulse which tune the dynamica phases such that $\phi_{11} = \phi_{01} + \phi_{10} - (2n + 1)\pi$, with integer n. With the help of U_{Cph} and qubit rotations, which form a set of universal gates, entanglement-on-demand and quantum algorithms can be implemented in real time.

A more demanding 10-qubit entanglement and parallel logic operations can be performed with a superconducting processor (Song et al. 2017). The processor has 10 transmons (Q_j, $j = 1 - 10$), having resonant frequencies tunable in the range 5–6 GHz and coupled to a central resonator, with fixed frequency of 5.795 GHz. Each qubit (transmon) has an individual microwave line, and an individual readout resonator. The processor is based on $\sqrt{iSWAP}$ gates and qubit rotations and is able to generate 10-qubit Greenberger-Horne-Zeilinger entangled states.

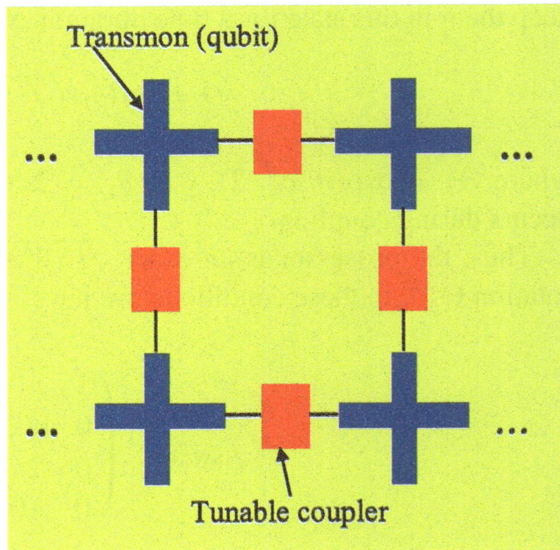

Fig. 4.18 Unit cell of the Sycamore programmable processor

The quantum computers working today at very big companies rely on building extremely complex superconducting microwave processors based on transmons. An example in this respect is the Sycamore programmable superconducting processor developed by Google, which has 53 transmon qubits, each qubit being coupled with other four in a rectangular lattice (Arute et al. 2019). The extreme difficulty of the design, fabrication and operation of such a quantum computer is suggested by the fact that the bibliographical reference (Arute et al. 2019) has 77 authors. The basic cell of this processor is represented in Fig. 4.18. Fast-fidelity gates that can be used simultaneously on the bidimensional array were developed, resulting in reduced errors.

The Sycamore processor has reached quantum supremacy, solving in 200 s a problem named random circuit sampling, which can be solved by a classical computer in 10,000 years. Quantum supremacy for a quantum computer means to solve in a short time certain algorithms, which are difficult to be solved by classical computers. However, a strong debate is ongoing about the meaning of quantum supremacy, and the time of computation for a classical computer to solve the above-mentioned problem. The debate is timely, since the era of quantum computers has just begin and quantum computers are notoriously difficult to be built, not to mention that the cost of their exploitation is very high. In fact, quantum computers incorporate the most advanced nanoelectronic devices to date.

Apart from superconducting qubits, many types of qubits and quantum gates based on semiconductor nanodevices have been proposed and demonstrated experimentally. For instance, we have shown in Chap. 3 that a quantum dot is an artificial atom, where the energy levels are discrete and equally distanced. Therefore, quantum dots are key nanodevices for quantum computing. A material of choice for the implementation of qubits in semiconductor nanodevices is Si, since it has a weak spin-orbit coupling and its isotopes have no nuclear spin, being a very pure crystal (Zwanenburg et al. 2013). In addition, Si is the basic material for nanoelectronic circuits, which are widespread in daily live, and its processing technology is developed in advanced foundries. There are also other materials used for semiconducting qubits, such as GaAs/AlGaAs heterostructures.

There are two large classes of semiconductor qubits that can be implemented with quantum dots: charge qubits and spin qubits. They are working at milliKelvin temperatures, as the transmons, but the hope is to reach 2–4 K as operating temperature, which could simplify the cooling equipments.

The charge qubit is based on two quantum dots, or an even number of quantum dots (Zhang et al. 2018). The qubit is simply defined as the surplus of electrons on the left or right dot, with electrochemical potentials E_L and E_R, respectively. We define the basis state $|0\rangle$ and $|1\rangle$, or (0,1) and (1,0), the surplus of electrons in the right and left dots, respectively. By denoting the detuning as $\Delta = E_R - E_L$ and the anti-crossing gap due to interdot tunneling as 2γ, we get the Hamiltonian

$$H_C = (\Delta\sigma_z/2) + \gamma\sigma_x \tag{4.28}$$

In fact, the potentials in the left and right dots are tuned by gates having applied voltages V_L and V_R, respectively, in terms of which a stability diagram and the energy level spectrum can be represented, as shown in Fig. 4.19 (Zhang et al. 2018). The arrow in the stability diagram marks a change in the qubit state. By tuning the Δ parameter, crossing and anti-crossing paths form. At $\Delta = 0$, coherent oscillations with frequency $\omega = \sqrt{\Delta^2 + (2\gamma)^2}/\hbar$ are observed, whereas by tuning Δ the charge qubits rotate on the Bloch sphere.

For a better understanding of the semiconducting qubit, we present below an example among many others. Excellent reviews on this topic are (Zwanenburg et al. 2013; Zhang et al. 2018). In this charge qubit based on the GaAs/AlGaAs two-dimensional gas (Cao et al. 2013), the excess of valence electrons in the left and right of two coupled quantum dots are denoted as the basis states $|L\rangle$ and $|R\rangle$ one charge

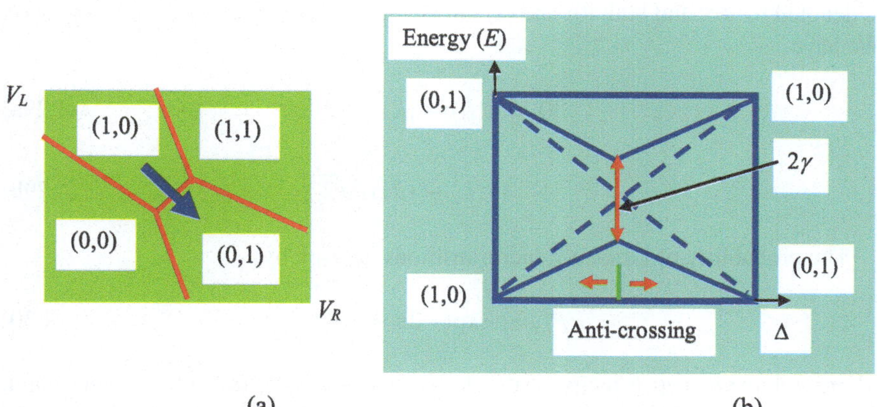

Fig. 4.19 Stability (a) and energy diagrams (b) of charge qubits

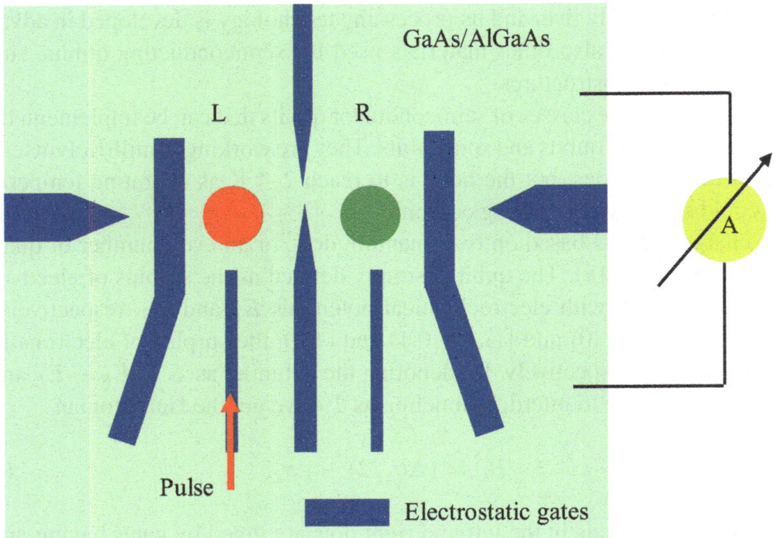

Fig. 4.20 Double quantum dot defined by electrostatic gates acting as charge qubit

qubit can be implemented using SiGe/Si double qunatum dots driven by microwave signals (Kim et al. 2015). The double quantum dot is presented schematically in Fig. 4.20. The interdot tunneling, which is coupling the left and right dots, displays an anti-crossing gap of $2\gamma = 20.7$ meV, corresponding to a Rabi frequency of 5 GHz, which was experimentally determined from the coherent oscillation produced by a square pulse. Each dot is occupied by 2–3 electrons. The system is initially in the $|R\rangle$ state, with a positive detuning Δ away from resonance. During the rising time of the pulse, the system is adiabatically driven to the anti-crossing point at $t = t_1$, where the excited state $|1\rangle$, with energy E_1, is reached (the ground state, with energy E_0, is denoted by $|0\rangle$, and both $|0\rangle$ and $|1\rangle$ are charge eigenstates away from resonance). We have

$$|0\rangle = (|L\rangle + |R\rangle)/\sqrt{2} \tag{4.29a}$$

$$|1\rangle = (|L\rangle - |R\rangle)/\sqrt{2} \tag{4.29b}$$

The probability of Landau-Zener transitions is given by

$$P_{LZ} = \exp(-2\pi\gamma^2/v\hbar) \tag{4.30}$$

where v is the sweep velocity of the driven pulse through the anti-crossing point. When the system is returning to $\Delta = 0$ at $t = t_2$, two paths develop due to coherent interference, with the phase difference

$$\delta\phi = (-1/\hbar)\int_{t_1}^{t_2}(E_1(t) - E_0(t))\mathrm{d}t \tag{4.31}$$

The readout is performed via the so-called Landau-Zener-Stückelberg (LZS) interference at the end of pulse, in the following way: $|L\rangle$ is determined by constructive interference, while $|R\rangle$ by destructive interference. The tunability of the charge qubits is achieved by changing the amplitude of the pulse. The measurements are performed at 30 mK.

Based on the above results, a CNOT gate was implemented on GaAs/AlGaAs (Li et al. 2015), while using quadruple Si/SiGe quantum dots a conditional phase gate was demonstrated (Ward et al. 2016). Both gates work at milliKelvin temperatures.

The spin qubit is another type of qubit that can be implemented in semiconductor nanostructures (Zhang et al. 2018). The basis state consists, now, of up and down spins, denoted as $|0\rangle$ and $|1\rangle$, which are split in a magnetic field B_0 applied along the z direction by the Zeeman energy $E_z = g\mu_B B_0$. The Hamiltonian is given by

$$H_{spin} = \hbar\omega_z\sigma_z/2 \tag{4.32}$$

where $\omega_z = g\mu_B B_0/\hbar$ is the Larmour rotation frequency of the spin around the z axis.

Additionally, electron spin resonance (ESR) or electric dipole spin resonance (EDSR) is needed to rotate the spin around the x axis of the Bloch sphere. In the case of ESR, an additional alternative magnetic field $B_{AC} \propto \cos(\omega t + \phi)$ must be applied along the x direction, with angular frequency ω, so that

$$H_{ESR} = \hbar(\omega_z - \omega)\sigma_z/2 + \hbar\omega_1(\sigma_x \cos\phi - \sigma_y \sin\phi) \tag{4.33}$$

where $\omega_1 = 2\pi f_R$, with $f_R = g\mu_B B_{AC}/4\pi$ the Rabi frequency, associated to spin rotation around the x axis when $\phi = 0$.

Magnetic field pulses with tunable durations and phases are able to control the spin rotation axis and phase. In this respect, using ESR pulses, a MOS circuit containing a quantum dot and a SET was fabricated. The SET, placed near the quantum dot, acts as a sensor that monitors the electrons inside the dot. A transmission line is then used to manipulate the spin state (Veldhorst et al. 2014). The principle of the device is described in Fig. 4.21.

The device in Fig. 4.21 is fabricated on purified ^{28}Si, with very reduced concentration of ^{29}Si. The tunnel coupling time between the quantum dot and the reservoir is controlled by a gate to be 100 μs when the pulse is read. Microwave pulses are applied to a transmission line to create the B_{AC} field, driving thus transitions between the spin-down, $|\downarrow\rangle$, and spin-up states, $|\uparrow\rangle$, of the quantum dot. At $B_0 = 1.4$ T, the resonance frequency is $\nu_0 = (g\mu_B/h)B = 39.104$ GHz so that $g = 1.998$. The spin-up qubit shows coherent oscillations, which are not attenuated, and which can be described by $A(\omega_R/\omega_{tot})^2 \sin^2(\omega_{tot}\tau/2)$ having the visibility factor $A = 0.7$ over 30 μs duration. Here ω_R is the Rabi frequency and ω_{tot} is the total Rabi frequency.

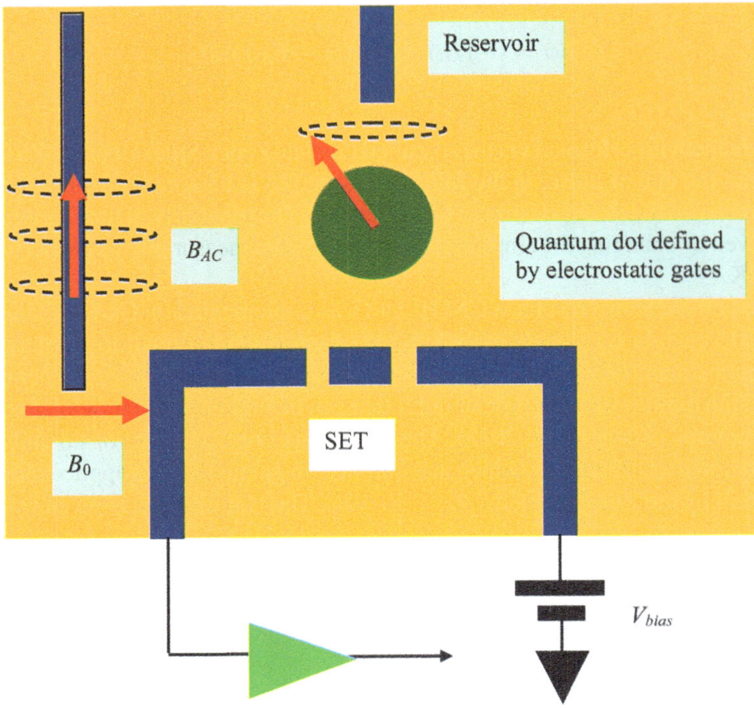

Fig. 4.21 The spin qubit in a quantum dot

The Rabi oscillations are varied by varying the control pulse applied on a control gate, which defines the quantum dot. The frequency of Rabi oscillations have the dependence $f_R = \omega_R/2\pi \approx B_{AC} \approx P_{ESR}^{1/2}$, where P_{ESR} is the rather-weak applied microwave power, around −20 dBm. By detuning the frequency, Ramsey fringes are produced when the spins are in the $x - y$ plane of the Bloch sphere, and the decay time T_2 is 120 μs. By removing various noise effects using echo-techniques, an enhancement of the spin coherence time up to 1.2 ms is obtained. The fidelity of the quantum dot qubit is of 99.59%.

The schematic charge stability diagram of spin qubits, i.e., the number of charges in the quantum dot, is represented in Fig. 4.22(a), and depends on the voltage on the control electrode and on the gate voltage where the microwave pulse is applied. Figure 4.22(b) displays the energy diagram, emphasizing the Zeeman splitting. The pulses required to control and readout the spin quantum dot are represented in Fig. 4.23.

The electric dipole spin resonance (EDSR) technique is implemented using two solutions. Both of them produce spin rotation using an oscillating electric field. The first solution is to place a gate above the quantum dot. If the quantum dot is made from a material with high spin-orbit coupling, the RF field applied on the gate is

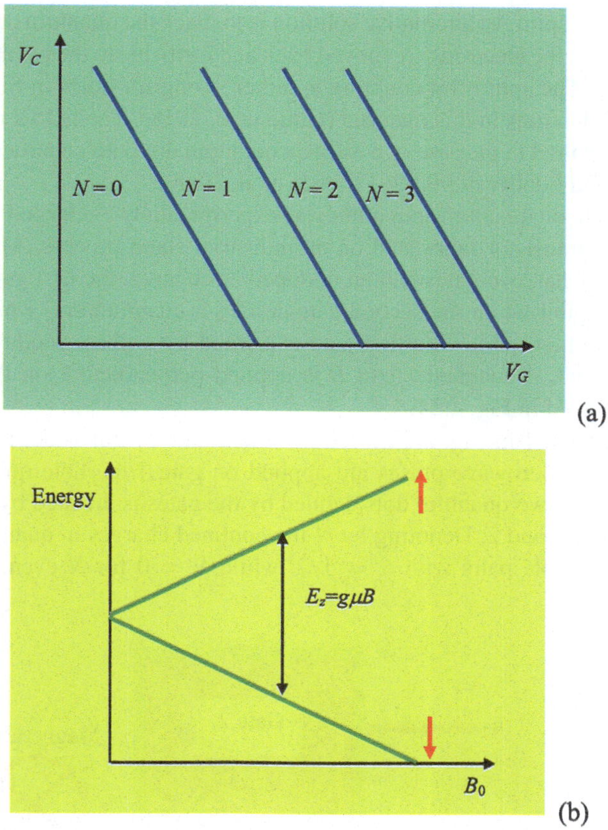

Fig. 4.22 (a) Charge diagram of the spin qubit and (b) energy level spectrum

Fig. 4.23 The control and the readout processes of the spin quantum gate

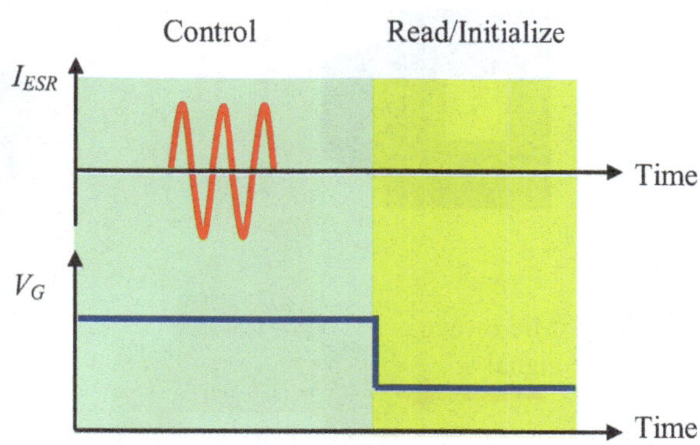

able to rotate the spin. An alternative solution is to place the quantum dot in a micro-magnet. As such, the electrons are moved back and forth due to the oscillations of the magnetic field. The spin-orbit coupling is rather strong for holes in Si and is weak in the case of electrons in this material (Hutin et al. 2018). The EDSR micromagnet solution was applied in the case of a Si/SiGe quantum dot with coherence limited by charge noise and fidelity of 99.9% (Yoneda et al. 2018).

A CMOS silicon quantum spin qubit was experimentally evidenced in a two-gate *p*-type Si FET transistor fabricated on an industrial stand process (Maurand et al. 2016). The FET has two gates and an undoped Si channel, the first gate having the role of a hole qubit while the second gate defines a quantum dot, which reads out the state of the first qubit. The first gate is excited by a phase modulated tunable microwave signal. A magnetic field B is applied perpendicular on the chip. The device is displayed in Fig. 4.24.

The FET channel in Fig. 4.24 is 10 nm thick and 20 nm wide, and each gate is 30 nm wide. Microwave pulses are applied on gate 1/the hole qubit. The hole occupation of the two quantum dots defined by the gates is dictated by the voltages applied on gates 1 and 2. Denoting by N the confined charges in quantum dots, for N odd we have hole pairs with $S = 1/2$, while $S = 0$ for N even. Thus, for N

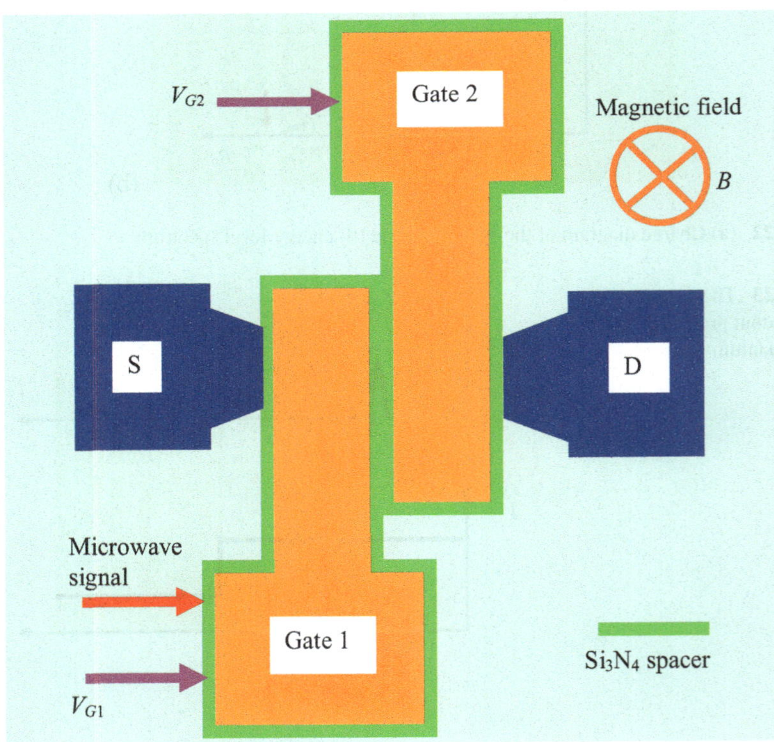

Fig. 4.24 The CMOS qubit FET

Table 4.3 The meaning of Pauli spin blockade

Quantum dot 1 spin orientation	Quantum dot 2 spin orientation	Tunneling between the two dots
↓	↑	Allowed
↑	↑	Not allowed

odd we have in both quantum dots holes with spin 1/2, i.e., the system is in the (1,1) charge configuration. If we require a transition (1,1) → (0,2), the Pauli spin blockade mechanism occurs. The meaning of Pauli spin blockade is illustrated in Table 4.3.

The spin blockade is controlled by suitable biases on the two gates and by microwave pulses. The EDSR takes place here by modulation of the continuous wave microwave signal, which rotates the spin in gate 1. In addition, controlled spin rotation can be achieved by microwave bursts, of duration τ_{MW}, instead of continuous microwave signals. In this last case, we have

$$\theta \approx \sqrt{P_{MW}}\tau_{MW} \tag{4.34}$$

where θ is the spin rotation angle, and P_{MW} is the power of the microwave burst. Rabi oscillations are present in the drain current dependence on the microwave pulse duration and are evidenced at detuned frequencies from the resonance, which is 9.68 GHz. The Rabi frequency, which increases also as $\sqrt{P_{MW}}$, can be as high as 85 MHz. Ramsey fringes are experimentally found to determine the dephasing time as 60 ns, the dephasing occurring due to the existence of Si^{29} nuclear spins inside the device and of other impurities.

There are few results concerning multi-qubit logic gates in Si. An example is a two-qubit gate fabricated on ^{28}Si epilayer, having very low residues of ^{29}Si, which is able to implement a CNOT gate using the ESR technique (Veldhorst et al. 2015). The implementation of this gate requires the controlled coupling of two quantum qubits. Each qubit is implemented by the spin of a single electron with energies split by a static magnetic field **B** of 1.4 T. The single-qubit state is controlled by ESR via a microstrip line integrated on the chip. The Stark shift is used to control electrically the spin resonance: $\nu = E_z/h$, where $E_z = g\mu_B B$. The device is not simple: it has three Al layers, nine Al gates, source, drain and reservoir electrodes fabricated using multiple stacks in CMOS technology. A SET is monitoring the charge state and the readout is performed via spin-charge conversion. A simple schematic cross-section of the 2-qubit spin quantum gate is displayed in Fig. 4.25.

The quantum device can work as a single dot or as a double quantum dot, depending on the biases applied on the four gates, which induce confining electrostatic potentials beneath G_1 and G_2. Another control gate G_C (not shown in Fig. 4.25) is placed bellow gates G_1–G_3 and spatially constrains the electron in the quantum dots along all sides, with the exception of the reservoir R. The coupling between quantum dots is controlled by electric fields, more precisely by energy detuning. The qubit Q1 is defined by loading one electron into it, a similar procedure being required for defining qubit Q2. The state of the device can be characterized by the numbers

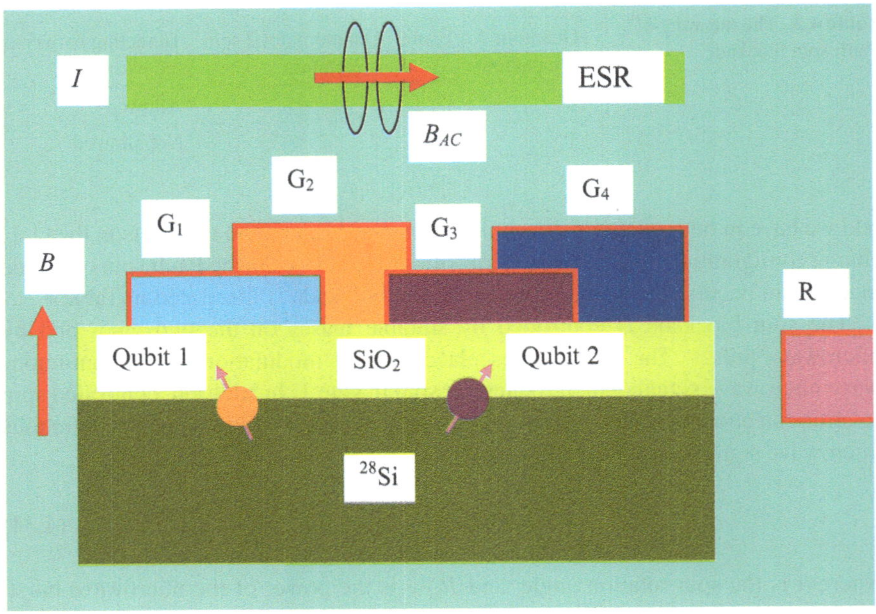

Fig. 4.25 Schematic cross-section of the two-qubit logic gate

Ni of electrons in the qubits Qi, $i = 1, 2$, as $(N2, N1)$. The device is controlled in the (1,1) state and the readout of Q1 is associated with the transition (0,1) → (0,0), while the readout of Q2 is related to the transition (1,1) → (0,1). If microwave pulses are applied on G1 and Rabi oscillations are measured, one can determine that Q1 has a dephasing time of 120 μs and a coherence time up to 20 ms. A controlled Z (CZ) gate can be obtained by detuning the energy of the quantum dots, such that the phase $\phi_{\uparrow\downarrow} + \phi_{\downarrow\uparrow} = \pi$; the CZ gate applies the Z gate, expressed by the Pauli matrix σ_z on a target bit, i.e. it applies a rotation of π to the target bit when its value is 1 when the control bit takes the logical value 1. This gate is thus equivalent to $C_{ph}(\pi)$. The CNOT gate can then be implemented by a rotation of the CZ gate, i.e., by applying two $\pi/2$ microwave pulses on the microstrip line.

A programmable two-qubit quantum processor was demonstrated also in Si (Watson et al. 2018) and was able to run Deutsch-Josza and Grover algorithms, which work must faster than classical algorithms on a classical computer, being based on quantum mechanical principles. In this case, a double quantum dot in Si/SiGe, as well as micromagnets, were used. These results were obtained also with the help of CZ gates, reaching a fidelity around 80%.

All the devices described above are working at milliKelvins. We can now understand why all quantum computers, which are in use at all big companies, are using transmons based on Josephson junctions. The Josephson junctions are much simpler devices compared to quantum dots, they can be easily replicated using standard technological processes and can be easily integrated in a microwave processor.

References

Arute F et al (2019) Quantum supremacy using a programmable superconducting processor. Nature 574:505–511

Cao G, Li H-O, Tao T, Wang L, Zhou C, Xiao M, Guo G-C, Jiang H-W, Guo G-P (2013) Ultrafast universal quantum control of a quantum-dot charge qubit using Landau-Zener-Stückelberg interference. Nat Commun 4:1401

Clarke J, Wilhelm FK (2008) Superconducting quantum bits. Nature 453:1031–1042

Dewes A, Ong FR, Schmitt V, Lauro R, Boulant N, Bertet P, Vion D, Esteve D (2012) Characterization of two-transmon processor with individual single-shot qubit readout. Phys Rev Lett 108:057002

DiCarlo L, Chow JM, Gambetta JM, Bishop LV, Johnson BR, Schuster DI, Majer J, Blais A, Frunzio L, Girvin SM, Schoelkopf RJ (2009) Demonstration of two-qubit algorithms with a superconducting quantum processor. Nature 460:240–244

Esmaeilzadeh H, Blem E, Armant RSt, Sankaralingam K, Burger D (2013) Power challenges may end the multicore era. Commun ACM 56:93–102

Gambetta JM, Chow JM, Steffen M (2017) Building logical qubits in a superconducting quantum computing system. NPJ Quantum Inform 3:2

Houck AA, Koch J, Devoret MH, Girvin SM, Schoelkopf RJ (2009) Life after charge noise: recent results with transmon qubits. Quantum Inform Process 8:105–115

Hutin L, Bertrand B, Maurand R, Crippa A, Urdampilleta M, Kim YJ, Amisse A, Bohuslavskyi H, Bourdet L, Barraud S, Yann XJ, Niquet M, Sanquer M, Bäuerle C, Meunier T, de Franceschi S, Vinet M (2018) Si MOS technology for spin-based quantum computing. 48th European solid-state device research conference (ESSDERC), Dresden, Germany, pp 12–17

Kim D, Ward DR, Simmons CB, Gamble JK, Blume-Kohout R, Nielsen E, Savage DE, Lagally MG, Friesen M, Coppersmith SN, Eriksson MA (2015) Microwave-driven coherent operation of a semiconductor quantum dot charge qubit. Nat Nanotechnol 10:243–247

Krantz P, Kjaergaard M, Yan F, Orlando TP, Gustavsson S, Oliver WD (2019) A quantum engineer's guide to superconducting qubits. Appl Phys Rev 6:021318

Li HO, Cao G, Yu GD, Xiao M, Guo GC, Jiang HW, Guo GP (2015) Conditional rotation of two strongly coupled semiconductor charge qubits. Nat Commun 6:7681

Maurand R, Jehl X, Kotekar-Patil D, Corna A, Bohuslavskyi H, Laviéville R, Hutin L, Barraud S, Vinet M, Sanquer M, De Franceschi S (2016) A CMOS silicon spin qubit. Nat Commun 7:13575

Nielsen MA, Chuang IL (2010) Quantum computation and quantum information. Cambridge University Press

Song C, Xu K, Liu W, Yang C-P, Zheng S-B, Deng H, Xie Q, Huang K, Guo Q, Zhang L, Zhang P, Xu D, Zheng D, Zhu X, Wang H, Chen Y-A, Lu C-Y, Han S, Pan J-W (2017) 10-qubit entanglement and parallel logic operations with superconducting circuits. Phys Rev Lett 119:180511

Stolze J, Suter D (2004) Quantum computing. A short course from theory to experiment. Wiley-VCH

Veldhorst M, Hwang JC, Yang CH, Leenstra AW, de Ronde B, Dehollain JP, Muhonen JT, Hudson FE, Itoh KM, Morello A, Dzurak AS (2014) An addressable quantum dot qubit with fault-tolerant control-fidelity. Nat Nanotechnol 9:981–985

Veldhorst M, Yang CH, Hwang JC, Huang W, Dehollain JP, Muhonen JT, Simmons S, Laucht A, Hudson FE, Itoh KM, Morello A, Dzurak AS (2015) A two-qubit logic gate in silicon. Nature 526:410–414

Ward DR, Kim D, Savage DE, Lagally MG, Foote RH, Friesen M, Coppersmith SN, Eriksson MA (2016) State-conditional coherent charge qubit oscillations in a Si/SiGe quadruple quantum dot. NPJ Quantum Inform 2:16032

Watson TF, Philips SGJ, Kawakami E, Ward DR, Scarlino P, Veldhorst M, Savage DE, Lagally MG, Friesen M, Coppersmith SN, Eriksson MA, Vandersypen LMK (2018) A programmable two-qubit quantum processor was demonstrated in silicon. Nature 555:633–637

Yoneda J, Takeda K, Otsuka T, Nakajima T, Delbecq MR, Allison G, Honda T, Kodera T, Oda S, Hoshi Y, Usami N, Itoh KM, Tarucha S (2018) A quantum-dot spin qubit with coherence limited by change noise and fidelity higher than 99.9%. Nat Nanotechnol 13:102–106

Zhang X, Li H-O, Wang K, Cao G, Xiao M, Guo G-P (2018) Qubits based on semiconductor quantum dots. Chin Phys B 27:020305

Zwanenburg FA, Dzurak AS, Morello A, Simmons MY, Hollenberg LCL, Klimeck G, Rogge S, Coppersmith SN, Eriksson MA (2013) Silicon quantum electronics. Rev Mod Phys 85:961–1012

Chapter 5
Neuromorphic Computation

Abstract Neuromorphic computation is based on the analogies with human brain functioning. Artificial neurons and synapses are key devices for such type of computing, and are based on memristors. This type of computing is dedicated for artificial learning, and the operation of artificial organs, such as retinas and bionic eyes.

5.1 Brain-Like Computation

A computing machine is based on serial instructions in a central processing unit (CPU), containing an arithmetic logic unit, and some data are transferred to a distinct memory unit and fetched from there. All operations are synchronized by clock pulses. Present-day computers are based on digital logic, i.e., the *on/off* state of billions of transistors, which are integrated on the same chip. Very large computers, having hundreds of thousands of CPUs simulating a cat brain, are consuming 3 MW of electrical power. In contrast, the human brain has 10^{11} neurons and 10^{15} synapses (Kuzum et al. 2013) and consumes only 20 W (Nawrocki et al. 2016), data computation and processing being analogical, based on pulses or spikes. Each neuron has a body, termed as soma, which is terminated with axons, along which the pulses propagate from one neuron to another. A simple electronic analogue of a neuron is displayed in Fig. 5.1.

The output of the neuron can be mathematically expressed as

$$y = a \sum_{1}^{N} w_i x_i \tag{5.1}$$

Information is processed and memorized at the level of each neuron and massive parallel processing is taking place at the level of the neuronal network. Each neuron generates spikes, with the amplitude of 0.1 V and duration in the range 0.1–1 mS. A neuronal network configuration is displayed in Fig. 5.2. Each neuron is connected with other neurons via 10^3 to 10^4 synapses. Neuron computation is performed via integration of various signals and generation of new spikes.

M. Dragoman and D. Dragoman, *Atomic-Scale Electronics Beyond CMOS*,
https://doi.org/10.1007/978-3-030-60563-6_5

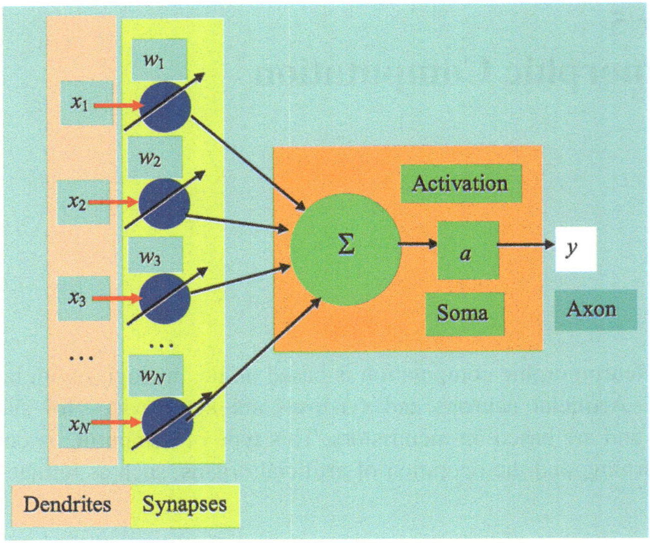

Fig. 5.1 Simple neuron model

Fig. 5.2 The unit cell of neuronal network

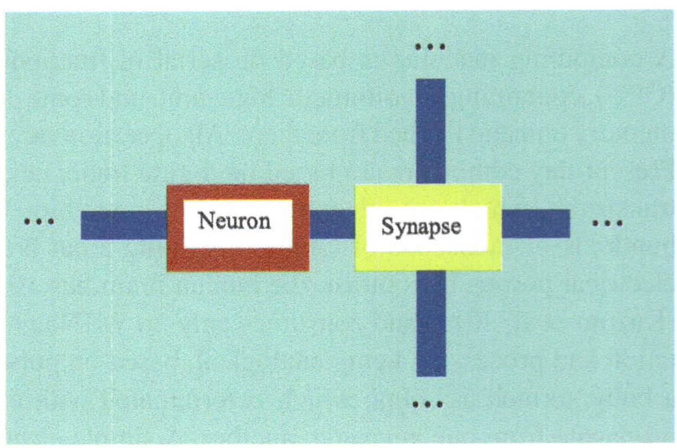

The synapses are gaps of 20–40 nm between axon ends and dendrites. They act as signal transmitters via chemical signals, when neurotransmitters are enabled, or via electrical signals. The dendrites are considered as input information channels, while axons constitute the path of the output information. The synapses are considered today the place where the learning and memory processes are taking place. In the neuron model formula (5.1) the role of synapses is represented by the weights w_i. These weights are variable in time, and the mechanism responsible for changing w_i, as well as for learning and memory, is termed synapse plasticity. The best known mechanism of changing the synapses' weights and hence of brain computation, is called spike-timing-dependent-plasticity (STDP). Note that SDTP is not the only process for computation in the brain. There is also the so-called Hebbian learning, which is widely used in neuromorphic computing, but SDTP remains the basic mechanism for this type of computation.

STDP tells us that, depending on the moment in time (before and after) of spikes generated by a pre-synaptic neuron and a post-synaptic neuron, w_i is increasing or decreasing. So, the timing of spikes is determining the function of the entire neuronal network.

We have represented schematically in Fig. 5.3 the STDP process, considering two neurons: the pre-synaptic neuron and the post-synaptic neuron. Each of them generates a pulse, referred to as "pre" and "post", the pulses being separated by a time interval Δt. Two situations are possible, as shown in Fig. 5.3:

- the pre-synaptic pulse is emitted before the post-synaptic pulse, $\Delta t < 0$, in a repetitive manner, case in which the synapse is potentiated and the associated weight or conductance is increased.
- the pre-synaptic pulse comes first and the post-synaptic pulse arrives after $\Delta t > 0$, case in which the synapse is depressed, i.e., its weight is decreased. The most encountered form of change in synaptic weight Δw (or conductance, ΔG, which is the measured parameter in electrical circuits) is represented as a function of time in Fig. 5.4.

This ordering in time of pulses governs the entire neuronal activity, and it was detected in many animals, not only in human beings.

Fig. 5.3 The SDTP process

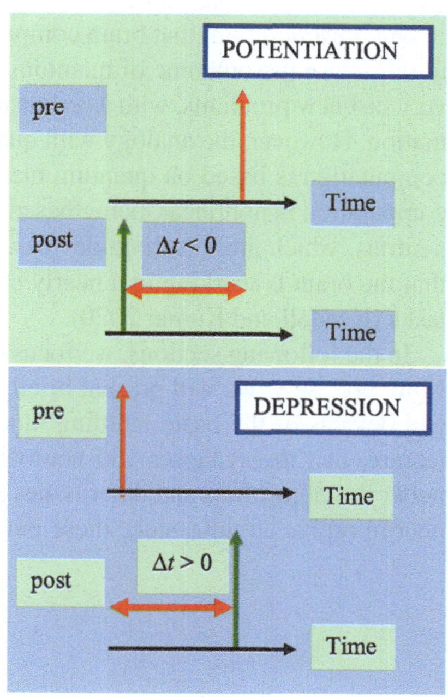

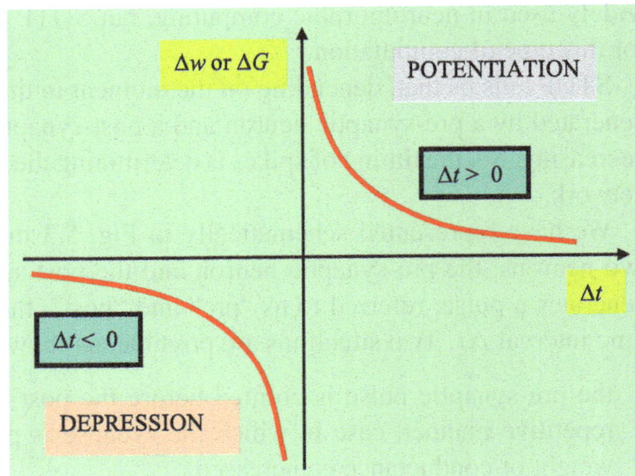

Fig. 5.4 SDTP weight/conductance change in time during potentiation and depression

There are few directions of development regarding neuromorphic computation: (i) artificial synapses and neurons implementation using new materials and new physical concepts, (ii) artificial neuronal networks, and (iii) development of a neuromorphic processor containing billions of transistors at a very-large-scale-of-integration, dedicated to learning vision, smell or other processes.

A neuronal network is formed from neuron layers, each layer being connected by synapses with pre- and post-synaptic neuron layers. The number of layers determines the depth of the neuronal network.

We must point out that brain computing is a probabilistic computing, resembling in this sense the outcome of quantum computing. In particular, the brain can solve very fast new problems, with a certain degree of probability, without too much information. However, the analogy with quantum computing stops at this level: quantum computation is based on quantum mechanics, which is linear, while neuromorphic computation is nonlinear, being based on nonlinear elements, such as synapses and neurons, which are threshold devices. In fact, based on recent researches, it seems that the brain is working in a nearly chaotic regime for learning and computational tasks (Kendall and Kumar 2020).

In the following sections, we focus on devices able to work as artificial synapses. More precisely, we will present briefly the implementation with different materials and devices of the basic building blocks of any neuromorphic computation architecture, i.e., the synapses and neurons. We will not tackle the topics of neuronal networks, algorithms and other issues related to the software or the hardware part of neuromorphic circuits, since these issues are beyond the aim of this book.

5.2 Synaptic Electronics

The first results on neuromorphic computation and artificial synapses and neurons were based on VLSI circuits (Mead 1989). In this seminal work, the basic artificial synapses and neuronal circuits can be found, and they constitute the foundation of present-day processors, which contain billions of transistors and which will be presented later. The tendency was to find single devices able to perform as a synapse, and many were found having two, three or multiple terminals. The main property of such devices is a controllable conductance in time, which can mimic SDTP. The memristor is considered today as the main candidate for an artificial or electronic synapse.

As an example, in the seminal paper of Jo et al. (2010), a non-uniformly Ag-doped Si vertical memristor was fabricated (see Fig. 5.5), containing an Ag-reach (high-conductivity) and an Ag-poor region (low-conductivity) produced by Ag doping. The device was excited by 100 positive pulses with the amplitude of +3.2 V and 300 μs duration, followed by 100 pulses with negative amplitude and 300 μs duration. Figure 5.6 illustrates the evolution of the measured current as a function of the number of pulses. This figure shows that the current is increasing as the number of positive, potentiating pulses is increased and decreases as the number of negative, depressing pulses is increased.

On the other hand, by applying positive and/or negative pulses with different durations, the conductance G changes in the following way: if the duration of the (positive or negative) pulse is increased, the conductance increases, while if the pulse duration is decreased the conductance decreases. Figure 5.7 illustrates the changes in conductance, ΔG, as a function of the pulse duration.

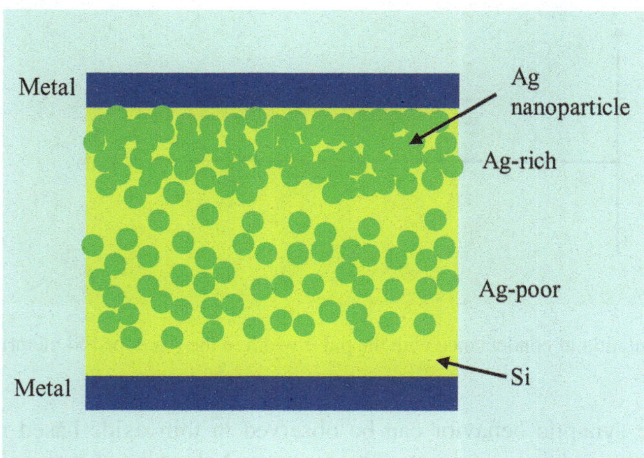

Fig. 5.5 The Ag-doped Si memristor

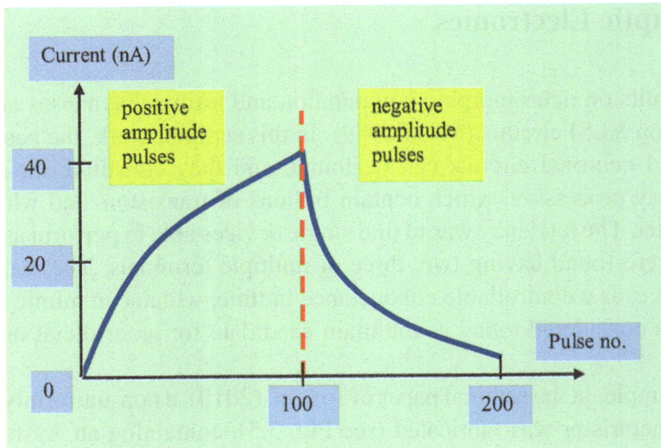

Fig. 5.6 Current as a function of the number of the pulses in the Ag-doped Si memristor

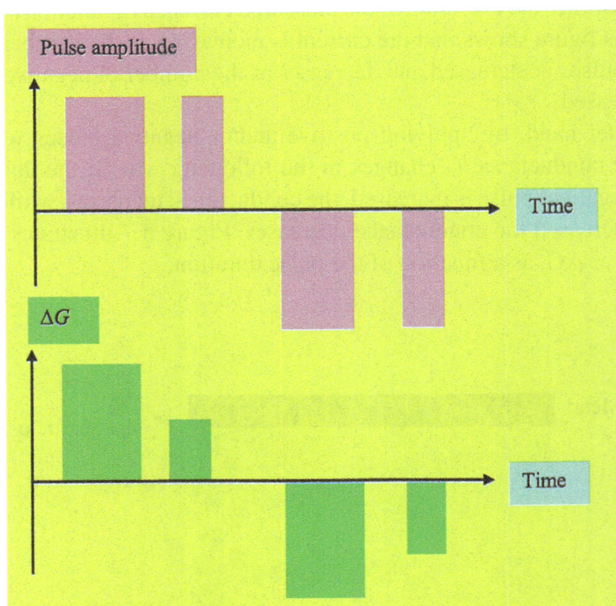

Fig. 5.7 Modulation of conductance with the pulse width in the Ag-doped Si memristor

The same synaptic behavior can be observed in thin oxide-based memristors, which constitute today the majority of memristors. A short list of such oxide memristors is found in Kuzum et al. (2013), while a more extended list can be discovered in Hong et al. (2018).

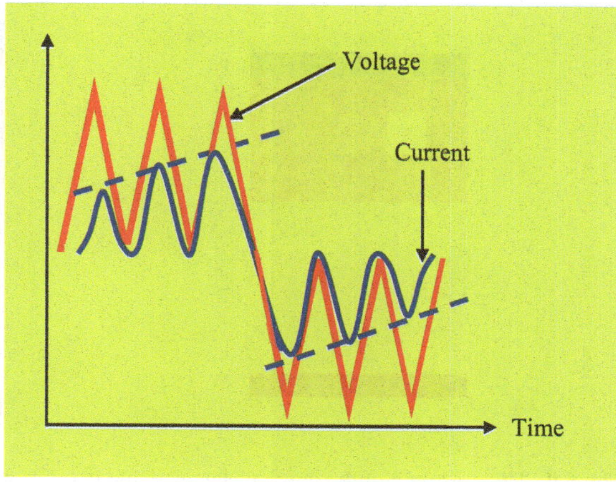

Fig. 5.8 Voltage and current dependences on time in an oxide memristor acting as synapse

An example of oxide synapses is WO_x, with the thickness of 50 nm, fabricated in a MIM structure (Chang et al. 2011). The synaptic behavior, i.e., the conductance modulation, was observed continuously as a result of the history of the applied voltage, e.g., positive or negative voltages that increase/decrease the conductance (see Fig. 5.8), or of the time interval between pulses.

The equations describing such a two-terminal oxide memristor behaving as a synapse are given by

$$i = G(w, v)v$$
$$\mathrm{d}w/\mathrm{d}t = f(w, v) \tag{5.2}$$

The first relation, between current i and voltage v, is valid in any resistive device but in this case it depends on an internal variable state w, while the second equation is the equation which defines the variable state.

A first hint about the w function was that it could be seen as an index describing the evolution of the lengths of the doped/undoped regions in the two-terminal memristor based on thin oxides, as indicated in Fig. 5.9, left. However, Chang et al. (2011) have elaborated a model in which w is determined from the drift of oxygen vacancies:

$$\frac{\mathrm{d}w}{\mathrm{d}t} \approx \sinh\left(\frac{V/E_0}{d - w}\right) \tag{5.3}$$

where E_0 is an electric field determined from experimental data, and d is the oxide thickness.

The model of w as conductive area is depicted in Fig. 5.9, right. There are two possible conduction mechanisms: (i) Schottky conduction, since the bottom electrode (W) forms a barrier with WO_x, and (ii) tunneling conduction, when filaments are formed. Thus, when a positive bias voltage is applied, the oxygen vacancies are transported by tunneling towards the top electrode, while negative bias voltages

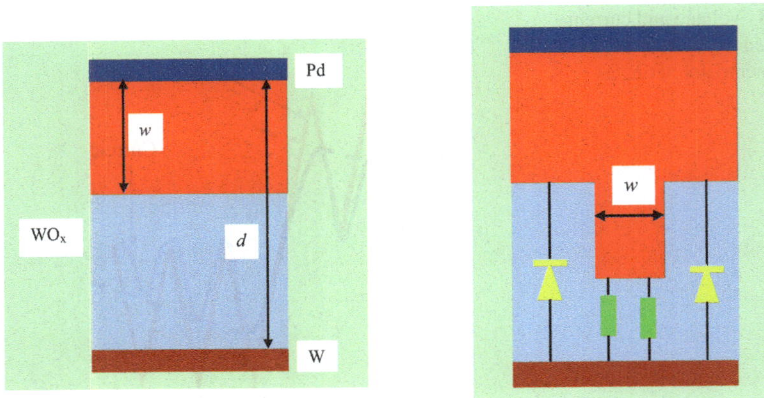

Fig. 5.9 The modeling of the internal state variable in oxide memristor

change the direction of current filaments towards the bottom electrode. In the absence of tunneling, the conduction is very low. The equations able to explain the synaptic behavior of the WO_x memristor are

$$I = \alpha(1 - w)[1 - \exp(-\beta V)] + w\gamma \sinh(\delta V) \tag{5.4a}$$

$$\mathrm{d}w/\mathrm{d}t = \lambda[\exp(\eta_1 V) - \exp(\eta_2 V)] \tag{5.4b}$$

The first equation is the $I - V$ relation in which the first term is due to the Schottky effect and the second is related to the tunneling effect. The second equation defines the memristor state variable: when $w = 0$, the Schottky transport mechanism prevails, whereas $w = 1$ is associated to the tunneling transport.

A more elaborate model (Chen et al. 2013) takes into account the forgetting effect observed in biological systems, i.e., a decay of conductance in time up to zero, as shown in Fig. 5.10. This model is based on the equations

Fig. 5.10 The evolution of conductance G in time associated to the forgetting effect

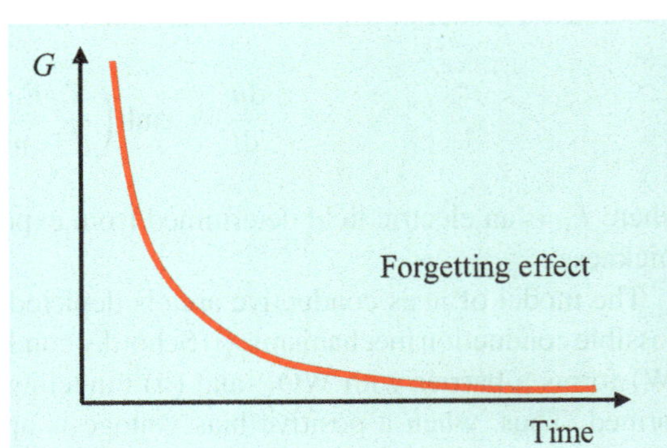

$$I = \alpha(1 - w)[1 - \exp(-\beta V) + w\gamma \sinh(\delta V) \tag{5.5a}$$

$$dw/dt = \lambda[\exp(\eta_1 V) - \exp(-\eta_2 V)]f(w) - w/\tau \tag{5.5b}$$

where

$$f(w) = \{[\text{sign}\,(V) + 1][\text{sign}\,(1 - w) + 1] + [\text{sign}\,(-V) + 1][\text{sign}\,(w) + 1]\} \tag{5.6}$$

The short term to long term memory is also evidenced by this model.

As suggested by the above-mentioned examples, there is no unique model for memristors, each memristor based on a certain physical effect needing its own model. Complicated software models are used to model the memristor behavior, as can be seen from books (Vourkas and Sirakoulis 2016) and handbooks recently published about memristors (Chua et al. 2020).

The memristors studied in Chap. 3, in particular the memristors based on phase-change materials (PCMs), are also used as synapses, and have their own models (Wang et al. 2017). The change in synaptic conductance G during a time interval Δt is described in this case by

$$\Delta G = \alpha \Delta t \exp\left(-\beta \frac{G - G_{\text{min}}}{G_{\text{max}} - G_{\text{min}}}\right) \tag{5.7}$$

where G_{min} and G_{max} are, respectively, the minimum and maximum conductance values of PCM, and α and β are fitting parameters depending on the PCM material.

Ferroelectric devices, such as capacitors, tunneling junctions, or FETs, are as well memristors and synapses (Oh et al. 2019). The origin of the change in time of their electrical parameters is the variation in time of polarization:

$$P(t) = P_S\left\{1 - 2\exp\left[-\frac{t}{t_S}\exp\left(-\frac{E_a}{V/d_{FE}}\right)^n\right]\right\} \tag{5.8}$$

where P_S is the spontaneous polarization, E_a the activation electric field, t_S is the switching time, d_{FE} is the thickness of the ferroelectric film and the constant n is related to the nucleation mechanism.

The switching in time in ferroelectrics under the action of a switching current is produced by domain switching and consists of two contributions: a non-switching and a pure polarization switching contributions. Partial switching is obtained when the pulse width t_w is shorter than a threshold switching time, denoted by t_{th}. The polarization level, i.e., the remanent polarization P_r, is determined by both pulse amplitude and duration, as can be seen from Fig. 5.11. Thus, the main parameter of a ferroelectric, P_r, which represents the value of polarization at 0 V, is tunable in time.

For instance, in a simple MFM capacitor where the ferroelectric is HfZrO with a thickness of 10 nm (see Fig. 5.12), the remanent polarization P_r is following in

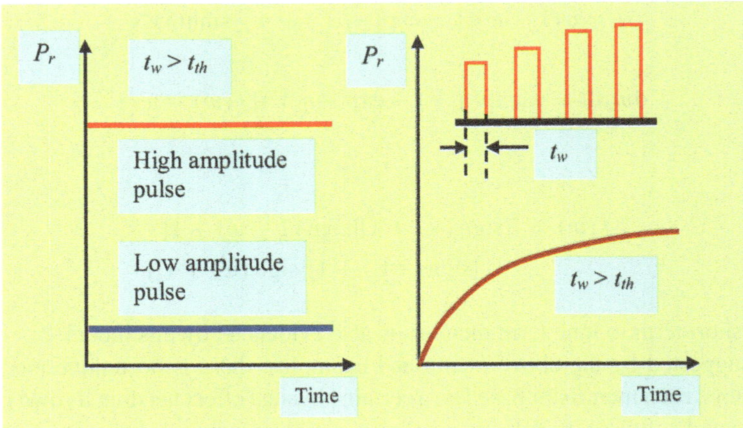

Fig. 5.11 Dependence of remanent polarization on pulse parameters

Fig. 5.12 HfZrO capacitor synapse

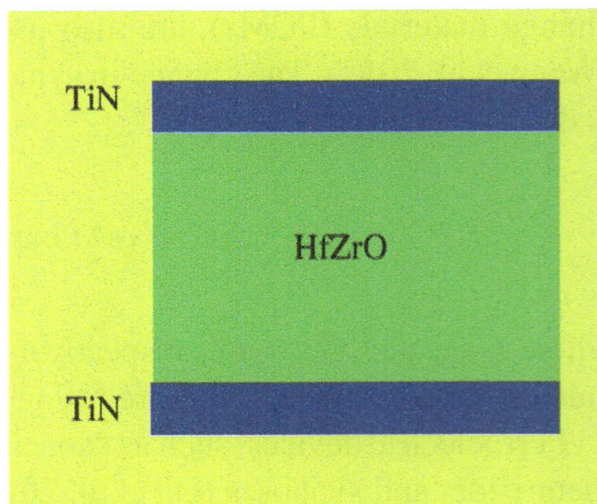

time the amplitude of the pulses that excite the capacitor. For positive amplitude pulses with duration of 50 μs, P_r is positive in time, while for negative amplitude pulses of the same duration P_r is negative (Oh et al. 2017), the remanent polarization maintaining its values even after a large number of switching cycles, as shown in Fig. 5.13.

There could be identified up to 32 levels of remanent polarization, which increase linearly with the pulse amplitude. To use these properties of the remanent polarization, the TiN/HfZrO/TiN capacitance is incorporated into a transistor as a gate capacitance. The device is called MFISFET. The drain current variation as a function of the number of pulses in this case is displayed in Fig. 5.14. The MFISFET with a doped Si channel works like a synapse due to the HfZrO ferroelectric layer. The same results were obtained for HfO_2 doped with Al, which has multilevel polarization values, so that an artificial neuronal network could be designed and simulated using such capacitors working as synapses (Zheng et al. 2019).

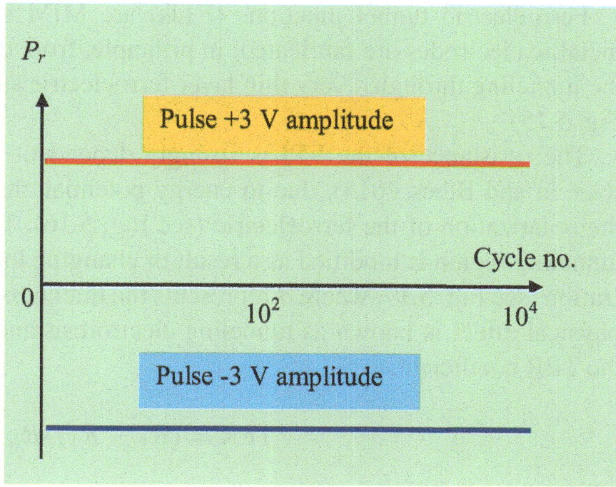

Fig. 5.13 Remanent polarization dependence on pulse amplitude as a function of switching cycle numbers

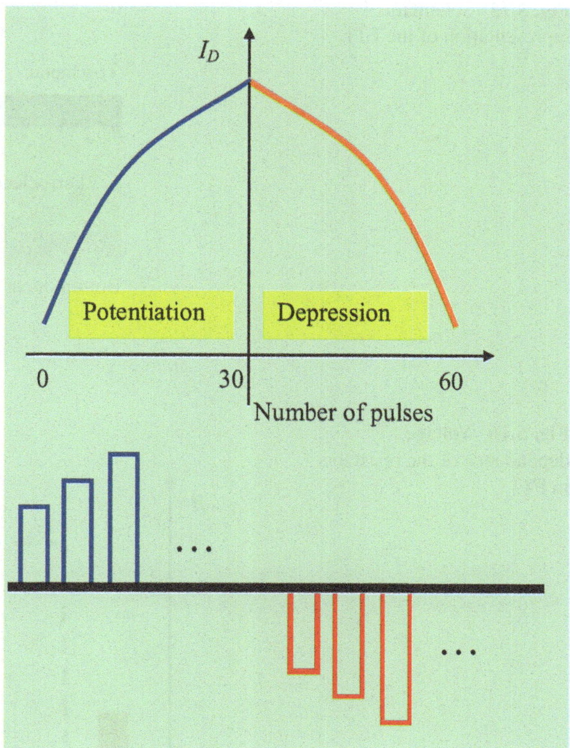

Fig. 5.14 MFISFET having TiN/HfZrO/TiN as gate capacitance

Ferroelectric tunnel junctions (FTJs) are MIM devices where top and down metallic electrodes are fabricated, in principle, from different metals for enhancing the tunneling through a very thin layer ferroelectric sandwiched between them (see Fig. 5.15).

The resistance of the FTJ is strongly dependent on the electrical polarization (Garcia and Bibes 2014), due to energy potential changes as a result of changing the polarization of the ferroelectric (see Fig. 5.16). Thus, the band diagram of the tunneling region is modified as a result of changing the direction of electrical polarization (see Fig. 5.17, where *d* represents the thickness of the potential barrier). This physical effect is known as tunneling electroresistance (TER) and is measured by the TER coefficient:

$$\mathrm{TER} = (R_{\uparrow} - R_{\downarrow})/R_{\downarrow} \tag{5.9}$$

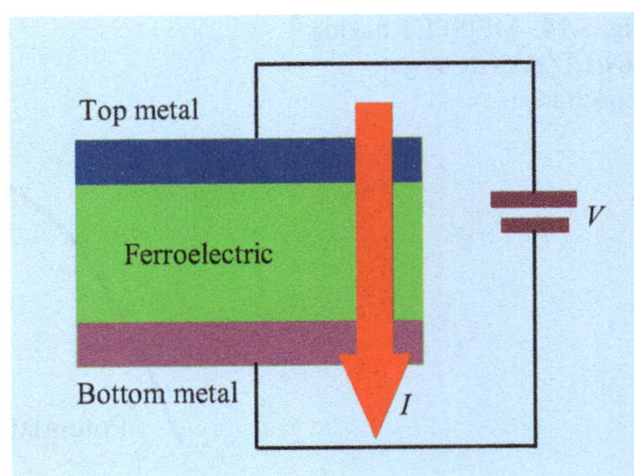

Fig. 5.15 Schematic representation of the FTJ

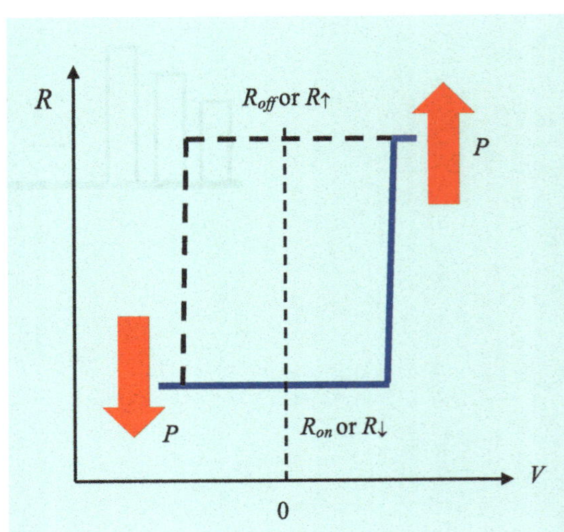

Fig. 5.16 Voltage dependence of the resistance in FTJ

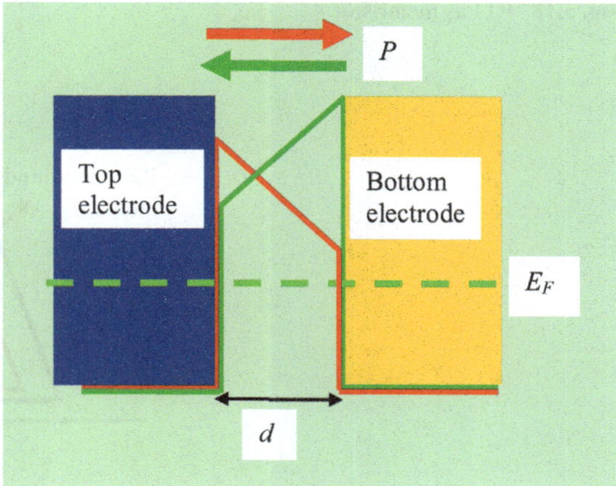

Fig. 5.17 The FTJ band diagram

Thus, RTJ is a resistive memory device, like many others described in Chap. 3. The FTJ was discovered many years before, but it was possible to be experimentally tested only when it was technologically possible to deposit few nanometers of ferroelectric. The introduction in FTJs and the history of this device are beautifully described in Guo et al. (2020).

If the carriers are travelling from the top electrode to the bottom electrode, they are tunneling a triangular barrier for one electric polarization, while when the polarization direction changes the height of the barrier increases and the current is very low. This is the TER effect. When the electrons are tunneling the triangular barrier with heights on the left and right sides denoted as φ_L and φ_R, the current has an analytical form given by

$$j(V) = -\frac{4em}{9\pi^2\hbar^3}\frac{\exp[\alpha(V)f(V)]}{\alpha^2(V)g(V)}\sinh[(3eV/4)\alpha(V)g(V)] \tag{5.10}$$

where

$$\alpha(V) = 4d\sqrt{2m}/[3\hbar(\varphi_L - \varphi_R + eV) \tag{5.11a}$$

$$f(V) = (\varphi_R - eV/2)^{3/2} - (\varphi_L + eV/2)^{3/2} \tag{5.11b}$$

$$g(V) = (\varphi_R - eV/2)^{1/2} - (\varphi_L + eV/2)^{1/2} \tag{5.11c}$$

and m is the effective mass of carriers.

The Fowler-Nordheim tunneling is a direct tunneling process as discussed above, at high voltages, while the thermionic charge transport appears whenever the carriers are transported by the applied field over the barrier, which is modeled as a Schottky

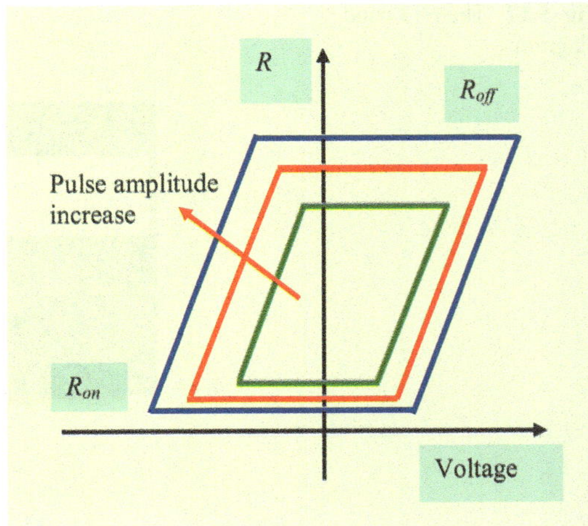

Fig. 5.18 FTJ as memristor

barrier. Both transport mechanisms may occur in a FTJ. There are many reviews about FTJ (see, for instance Garcia and Bibes 2014 and Guo et al. 2020).

If the voltage is swept several times, the resistance of the FTJ takes multiple values between R_{on} and R_{off} due to the dynamics of ferroelectric domain walls (nucleation and growth), and thus the FTJ behaves like a memristor (Chanthbouala et al. 2012). The dependence of the resistance of the FTJ for different amplitudes of the writing voltage pulses is displayed in Fig. 5.18.

The resistance of the FTJ can be modulated by pulses with different amplitudes, or by successive pulse trains with constant amplitudes. The multilevel response of the resistance is the imprint of the memristor and it can be expressed by the parameter s, which represents the relative fraction of domains with downwards polarization, given by

$$s = (1/R - 1/R_{on})/(1/R_{off} - 1/R_{on}) \tag{5.12}$$

where R_{on} and R_{off} correspond to upward and downward orientations of the polarization, respectively. In the *on* state we have $s = 0$, in the *off* state $s = 1$, otherwise s takes a value in the interval (0,1).

The equations of memristors based of FTJ can be written (Chanthbouala et al. 2012) as

$$v(t) = R(s, V, i)i(t) \tag{5.13a}$$

$$\mathrm{d}s/\mathrm{d}t = (1 - s)\frac{2}{\tau_P(V)}\frac{t - \tau_N(V)}{\tau_P(V)} \tag{5.13b}$$

where τ_N is the nucleation time and τ_P the propagation time of ferroelectric domains.

Many FTJs made from different ferroelectrics, such as BTO, BFO and HfZrO, display synaptic effects (see Table I from Guo et al. 2020). In principle, if the number of pulses increases, R_{on} decreases, whereas R_{off} increases with the pulse duration (Guo et al. 2018).

It is important to study FTJ devices acting as memristors and synapses using HfO_2-based ferroelectrics, which are compatible with CMOS technology, since in real neuromorphic computing systems an impressive number of synapses are needed to mimic the brain behavior. A simple FTJ of this type is the Pt/HfZrO/Pt heterostructure, where Pt is the metal of both top and bottom electrodes (Ambriz-Vargas et al. 2017). The band diagram of the heterostructure is displayed in Fig. 5.19.

The HfZrO ferroelectric layer has a thickness of 2.8 nm, and the current–voltage measurements were performed with a conductive AFM in direct contact with the top electrode (see Fig. 5.20), sweeping the voltage. The FTJ is conducting for downward polarization and is in the *off* state for upward polarization, as shown in Fig. 5.21.

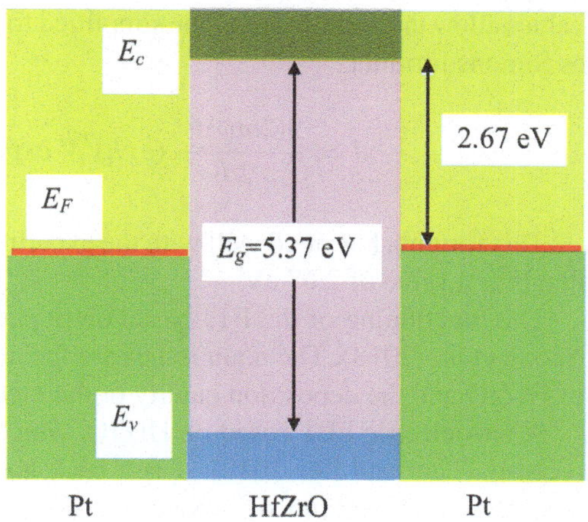

Fig. 5.19 The band diagram of Pt/HfZrO/Pt FTJ

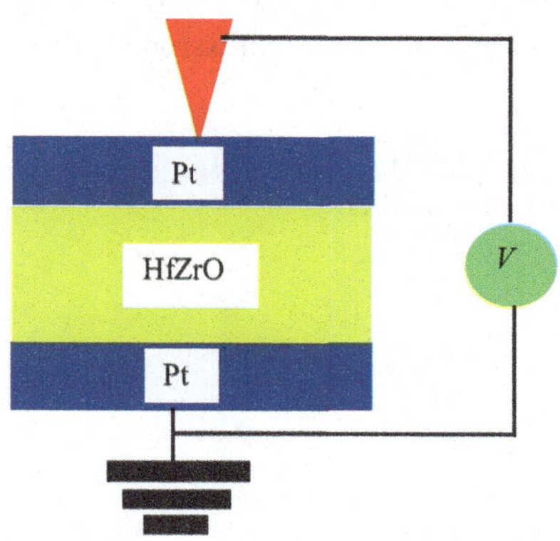

Fig. 5.20 The HfZrO FTJ structure

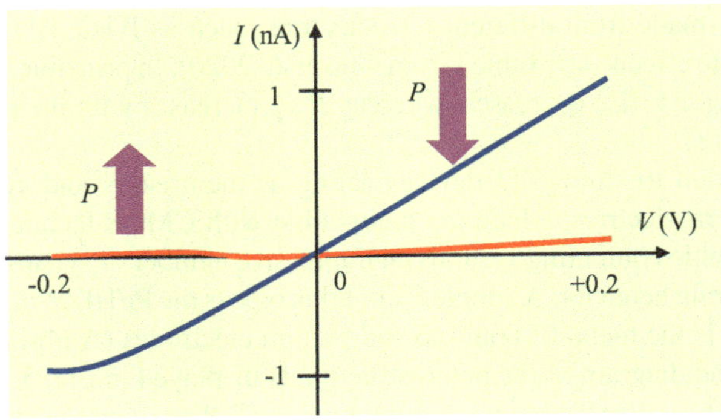

Fig. 5.21 Current–voltage dependence of the HfZrO FTJ (after Ambriz-Vargas et al. 2017)

From Fig. 5.21 it follows that the current is very low, around 1 nA, and the voltage range is narrow, since otherwise HfZrO could suffer breakdown. These low voltage values allow the application of the simplified formula of the tunneling current, known as Simons formula.

$$J = \frac{3(2m\varphi)^{1/2}}{2d}(e/h)^2 V \exp\left(-\frac{4\pi d(2m\varphi)^{1/2}}{h}\right) \tag{5.14}$$

This formula has fitted well with the experimental results, resulting for the barrier height φ a value of 2.67 eV.

The functioning of the FTJ based on doped HfZrO as a synapse is reviewed in Yoong et al. (2018). The main issues are the uniform deposition of few nanometers of HfZrO and the deposition quality of the metallic contacts.

A ferroelectric FET based on HfZrO, which behaves as a synapse, was reported recently (Kim and Lee 2019). It is a back-gate transistor configuration, having as channel indium gallium zinc oxide (IGZO), as indicated in Fig. 5.22.

The FET is a memory having a hysteretic behavior in the $I_D - V_G$ characteristic and multilevel behavior when the gate voltage is swept. The conductances in the potentiation and depression regimes are written as:

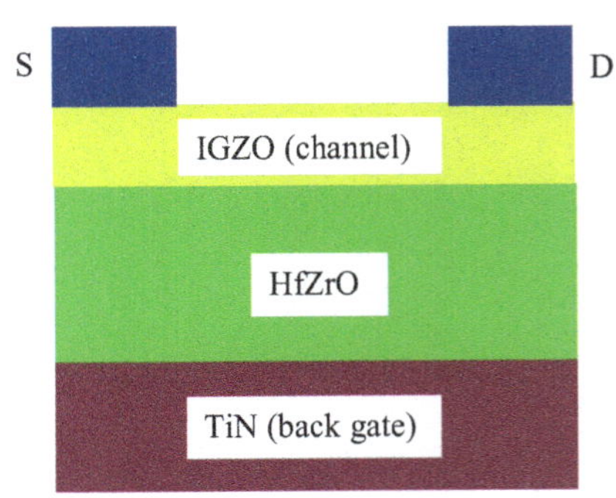

Fig. 5.22 The FET synapse based on HfZrO

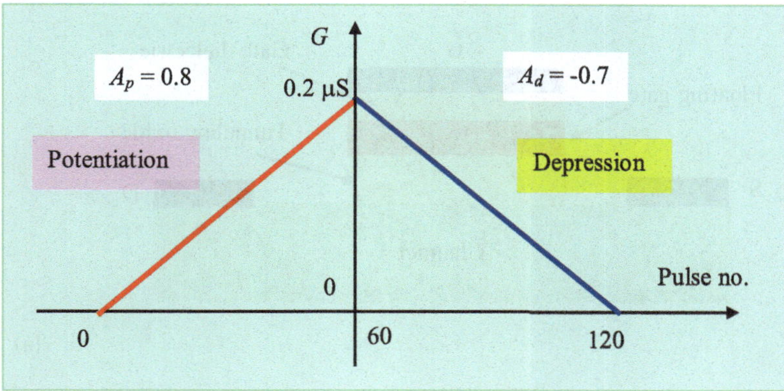

Fig. 5.23 Conductance dependence on the pulse number in a HfZrO FET (after Kim and Lee 2019)

$$G_p = B_p\left[1 - \exp\left(-\frac{p}{A_p}\right)\right] + G_{\min} \tag{5.15a}$$

$$G_d = -B_d\left[1 - \exp\left(\frac{p - p_{\min}}{A_d}\right)\right] + G_{\max} \tag{5.15b}$$

with

$$B_{p,d} = \frac{G_{\max} - G_{\min}}{1 - \exp(-p_{\max}/A_{p,d})} \tag{5.15c}$$

where G_p is the potentiation conductance, G_d the depression conductance, $A_{p,d}$ is the corresponding linearity factor, p is the pulse number and $p_{\max}$ the maximum number of pulses. The dependence of G of the number of pulses is represented in Fig. 5.23.

Memory FET transistors form also a large category of devices, which can be used as synapses and are fabricated using CMOS technology. We present very briefly this subject, since it was recently review in an excellent paper (Milo et al. 2020). The use of memory transistors as synapses has been shown first in Mead (1989).

FET memories are FETs with gate structures, which are equipped with additional layers, able to store and release charges. There are two solutions, depicted in Fig. 5.24, for such devices: a floating gate, which is a highly doped layer in the case of CMOS transistors, or a dielectric material, which traps the carriers and releases them in a controlled manner. Both have the role to shift the threshold voltage according to the formula

$$\Delta V_{\text{th}} = -Q_{\text{S}}/C_{\text{G}} \tag{5.16}$$

where V_{th} is the threshold voltage, Q_{S} is the stored charge and C_{G} is the gate capacitance. A negative shift of V_{th} means that positive charges are stored, while a positive

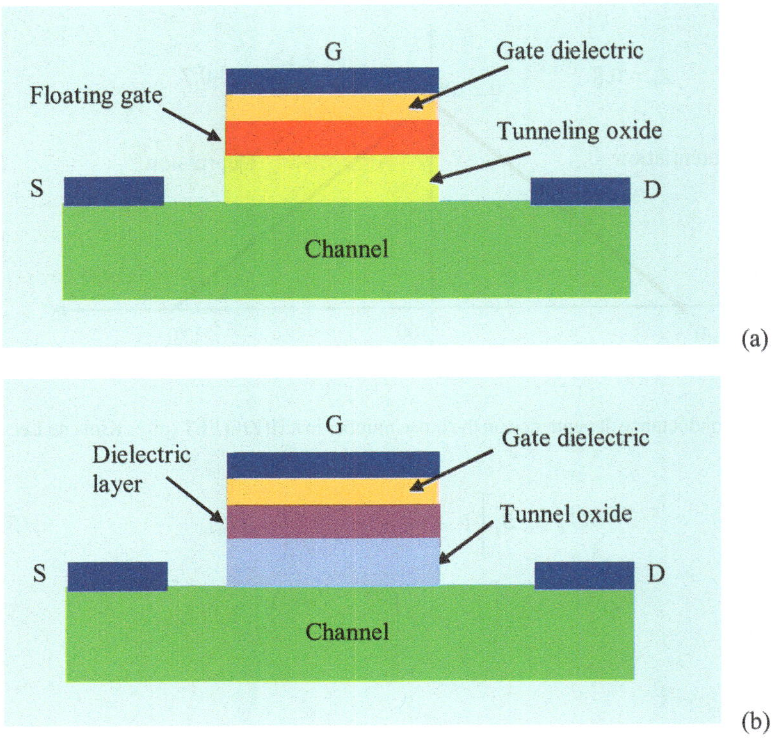

Fig. 5.24 (a) Floating gate and (b) dielectric layer for charge trapping in FET memories

shift of V_{th} indicates that negative charges are stored, as can be seen in the $I_{\text{D}}-V_{\text{G}}$ dependence illustrated in Fig. 5.25.

For programming and erasing of conductance states two physical mechanisms are used: (i) the Fowler-Nordheim tunneling between the substrate and the floating

Fig. 5.25 The working principle of memory FETs

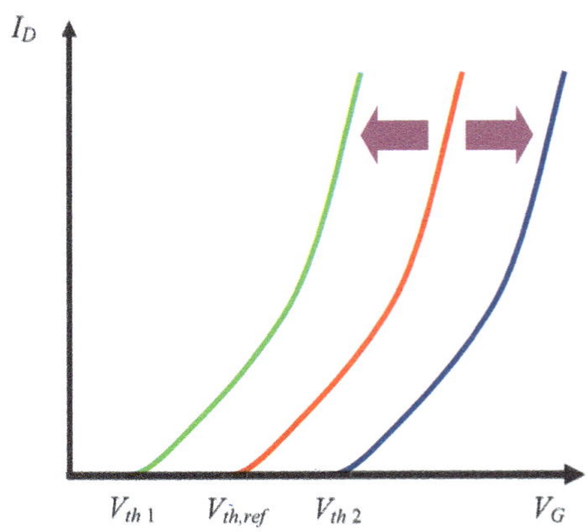

gate via the tunnel dielectric, by applying a high voltage on the channel, and (ii) channel hot electron injection (CHEI), which involves the application of a positive drain voltage, large enough for the electrons to overcome the tunneling barrier.

There are several memory architectures, such as NAND Flash or NOR Flash, and the advantage of memory transistors as synapses is that they can benefit of the state of the art of CMOS technology. More details are found in Milo et al. (2020).

The dependence of the drain current on the gate voltage in this case is given by

$$I_D = I_0 \exp[e\alpha(V_G - V_{th})/mkT] \exp(-e\alpha\Delta V_{th}/mkT) \tag{5.17}$$

where m denotes the subthreshold voltage slope, I_0 is a constant current, α is the gate-floating gate coupling capacitance ratio, and ΔV_{th} is the shift of the threshold voltage from a threshold voltage reference. The first exponential term in (5.17) can be seen as the pre-synaptic signal, while the second exponential term is the synapse weight:

$$\Delta w = \exp(-e\alpha\Delta V_{th}/mkT) \tag{5.18}$$

There are different changes in the configurations of standard nonvolatile memories that can be used as efficient synaptic analogues in the learning process (Milo et al. 2020). The ultimate synaptic devices or synaptic materials are based on atomic-thin materials and the devices based on them. A recent review in this area is (Sangwan and Hersam 2020). Excellent synaptic transistors are based on CNTs, which can be random (Shen et al. 2013) or aligned CNT arrays (Sanchez Esqueda et al. 2018), using the same principles mentioned above, i.e., floating gate or trapping dielectric layers. Other 2D materials have been used as synapses, such as MoS_2, graphene, or h-BN, but all the experiments based on them and reviewed in Sangwan and Hersam (2020) suffer from the lack of demonstration at the wafer scale. Since billions of synapses are required for efficient brain computing, the artificial synapses based on 2D materials are far away from a realistic implementation of neuromorphic computing.

The simplest artificial neural network is the perceptron, which can be implemented using various materials (see Fig. 5.26) in a configuration involving a crossbar array of memristors (Alibart et al. 2013).

The output of the binary perceptron can be expressed as

$$y = \text{sgn}\left[\sum_0^9 w_i x_i\right] \tag{5.19}$$

and the perceptron learning rule is given by

$$\Delta w_i = \alpha x_i (y_d - y_a) \tag{5.20}$$

where α is the learning rate, y_d is the desired output and y_a the actual output, while x_0 is the bias input.

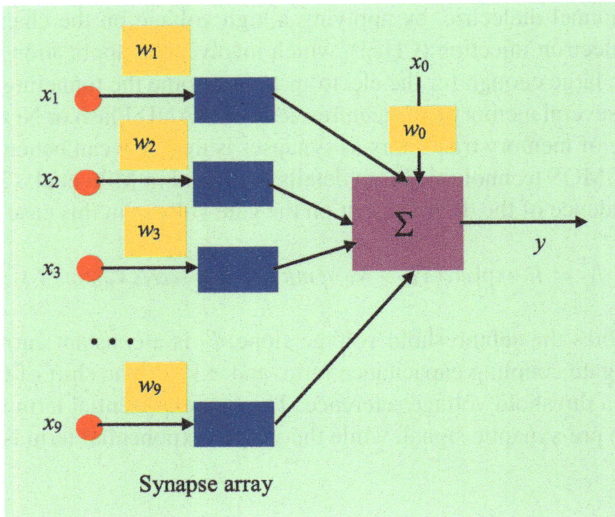

Fig. 5.26 The binary perceptron

There are nine binary inputs representing an array of 3×3 points. The bias input is $x_0 = +1$, and the input and output have values of $+1$ or -1. The role of this binary perceptron is to divide the input patterns into two distinct groups via the operation $\mathrm{sgn}\left[\sum_0^9 w_i x_i\right]$. The two patterns are the letters X and T consisting of 9 pixels formed from the inputs x_i having values of $+1$ or -1. An example is given in Fig. 5.27.

Practically, an input pattern is produced and the weights w_i are adjusted continuously for every new pattern, until the desired pattern is obtained. The training tests contain 50 patterns. One train epoch is performed by introducing randomly ordered patterns and updating w_i. After 17 epochs, the classification attains 100%.

Fig. 5.27 The letter T as a desired perceptron pattern

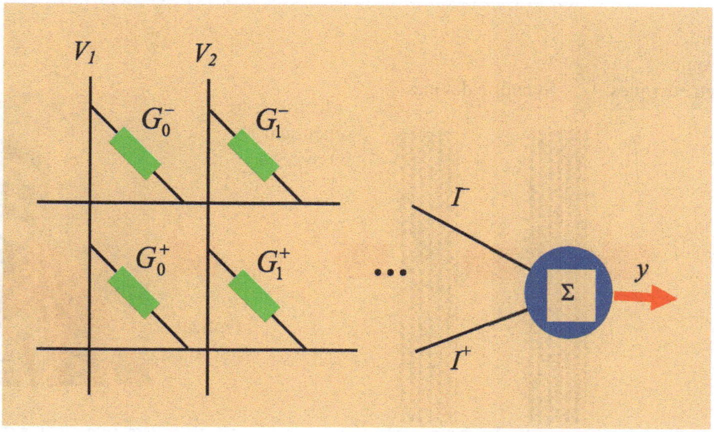

Fig. 5.28 The memristor perceptron

The synapses are implemented with an array of 2 × 10 MIM memristors of the type $Pt/TiO_{2-x}/Pt$, two memristors being assigned for every synapse, such that the synaptic weight is expressed by the memristor conductance $G_i = G_i^+ - G_i^-$. The memristors are arranged in a crossbar array (Fig. 5.28) where the inputs are voltages and the outputs are currents

$$I^+ = \sum_0^9 G_i^+ V_i \tag{5.21a}$$

$$I^- = \sum_0^9 G_i^- V_i \tag{5.21b}$$

Then, the perceptron output is given by

$$y = \text{sgn}[I^+ I^-] \tag{5.22}$$

These first attempts of pattern recognition based on memristor crossbar arrays (see Alibart et al. 2013 and the references therein) have important applications in many areas. One example refers to artificial retinas (Berco and Ang 2019), where the concept of computing in sensor is the crucial concept which can be fulfilled using crossbar arrays. An artificial retina is an electronic device able to mimic the retina functions in the eyes, capturing the optical images, processing them, and transforming the images into chemical signals to be sent to optical nerves. Artificial synaptic devices could play a key role, since the entire artificial retina could be seen as a combination of memristor crossbar arrays performing various tasks. The inputs are photodetectors that transform the visible light into electrical spikes, which are processed by several synaptic crossbar arrays, and then transformed into chemical

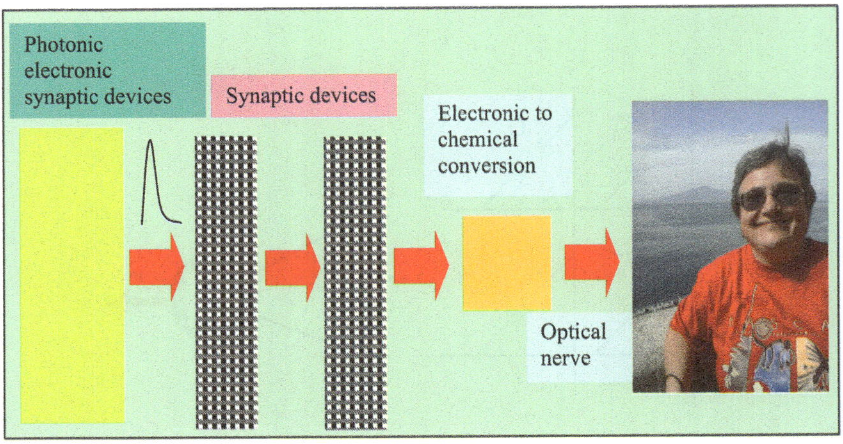

Fig. 5.29 The concept of the artificial retina chip

signals. The concept of an artificial retina is presented in Fig. 5.29 and the crossbar array implementation in Fig. 5.30.

The artificial retina is conceived as a 3D stack of crossbar arrays loaded with memristors, and the total dimension must not exceed 1 mm^3, implying that the synapses must be smaller than 20 nm and must consume energy at the level of few picoJoules. Of course, if the artificial retina is used for other applications, such as machine vision, or robots, the dimensions could be higher.

There are several categories of memristors able to react at optical stimuli. We have given an example in Chap. 3 of memristors reacting at visible light. There are many more such memistors based on organic materials, a large part of them being biocompatible.

If we look at the large table presented at the end of the review paper (Berco and Ang 2019) about the possible synaptic solutions for artificial retina, very few are possible good candidates for actually replacing the retinas existing in eyes. The actual implementations are rather big or consume too much current. A promising development is a $CH_3NH_3PbI_3$ (MaPbI3) memristor, which is an optogenetic inspired memristor (Zhu and Lu 2018). This memristor is changing the conductance from 15 μS to zero in 10–20 ms, depending on the density of power of the applied visible light. The power density of the visible light is very week, e.g., 0.2–0.3 $\mu W/cm^3$, and mimics the light-controlled Ca^+ ions in biological synapses.

Other types of biosensors are now using the concept of computing in sensors. An example in this respect is an artificial sensory nerve, which is made using organic electronics devices, such as pressure sensors and electronic circuits (see Fig. 5.31). In principle, an array of pressure sensors (1–8 kPa) produces variable voltages in time, which are transformed into a train of pulses by a ring oscillator working at 100 Hz. These are pre-synaptic pulses, which are collected by a transistor synapse (Kim et al. 2018). In this way, this electronic system acts as an artificial afferent nerve, which is able to detect the movement of objects and Braille characters. If this artificial nerve

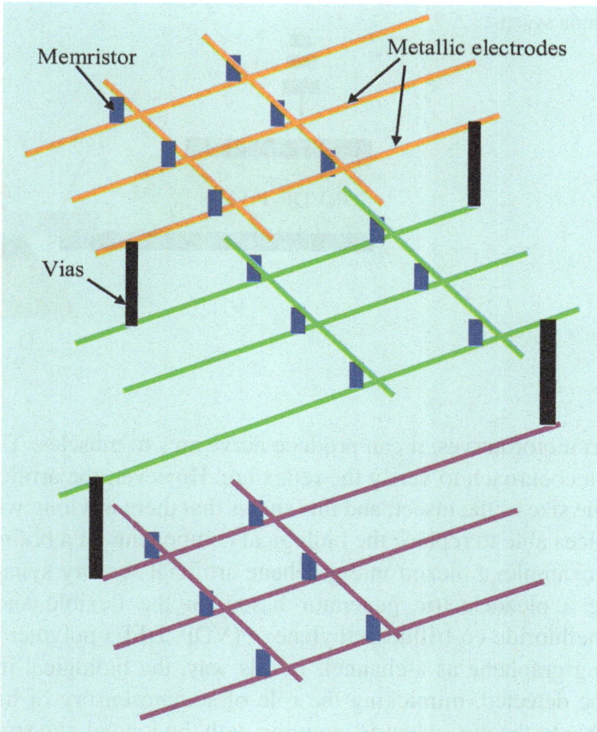

Fig. 5.30 Artificial retina as a network of crossbar arrays loaded with memristors (after Berco and Ang 2019)

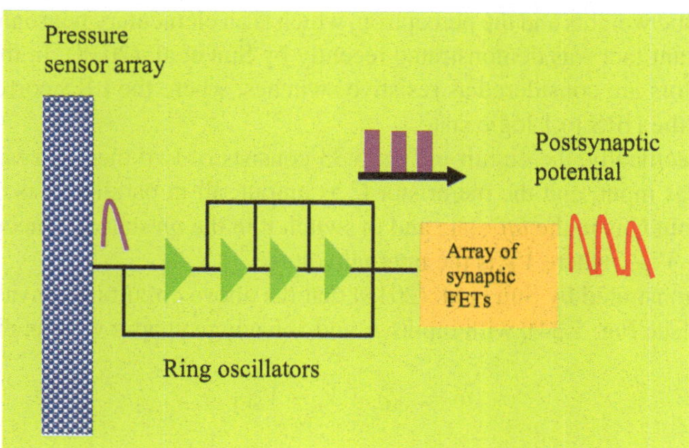

Fig. 5.31 The artificial nerve circuit

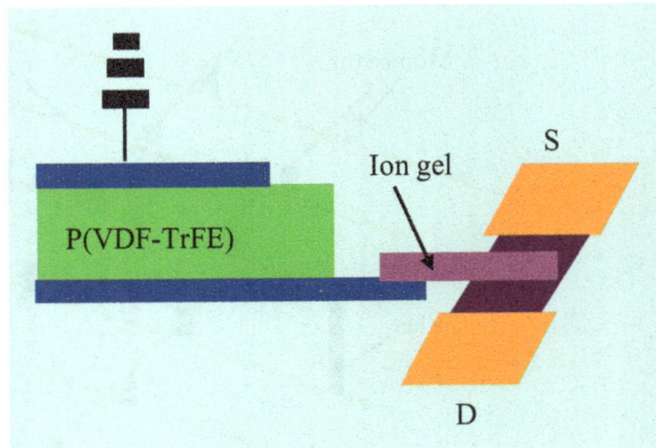

Fig. 5.32 The strain synapse sensor

is connected to motor nerves, it can produce nerve arcs to muscles. The entire chip is attached to a cockroach to verify the reflex arc. However, the artificial nerve has almost the same size as the insect, and this shows that there is a long way to produce electronic devices able to replace the biological components of a brain.

In another example, a piezotronic graphene artificial sensory synapse was built by integrating a piezoelectric generator based on the flexible and transparent poly(vinylidenefluoride-co-trifluoroethylene) P(VDF-TrFE) polymer with an ion-gel FET having graphene as a channel. In this way, the biological-inspired touch patterns can be detected, mimicking the role of somatosensory of humans (Chen et al. 2019). Due to the piezoelectric coupling with the ion gel, the strain amplitude and duration are transformed into synaptic signals via the FET (see Fig. 5.32).

Although neuromorphic computation is an analog computation, Boolean logic functions can be generated using the concepts presented up to now, i.e., the tunability of the synaptic weights and the perceptron, which is an elementary neuronal network. This important fact was demonstrated recently by Sun et al. (2018). In this respect, the memristors are considered as resistive switches, where the LRS corresponds to 0 logic and the HRS to 1 logic state.

More specifically, the circuit in Fig. 5.33 consists of two memristors, A and B, considered as input, and the memristor C as output, all in parallel. It is considered that the output C is in the *off* state and to switch it in the *on* state it is necessary that $V_C - V_P \geq V_{\mathrm{SET}}$, where V_P is the internal voltage.

It is demonstrated by Sun et al. (2018) that the above circuit is equivalent with a perceptron (see Fig. 5.34), with inputs A and B having synapse weighs given by

$$w_i = V_C - V_i - V_{\mathrm{SET}} \tag{5.23}$$

where $i = A, B$.

By tuning the w_A and w_B weights, various logic functions can be implemented, such as NAND, NOR and even other more complicated logic functions, such as XOR, or material implication IMP (Borghetti et al. 2010). The IMP logic gate (called also IMPLY) is considered to embody a 'stateful' logic operation, in which the same

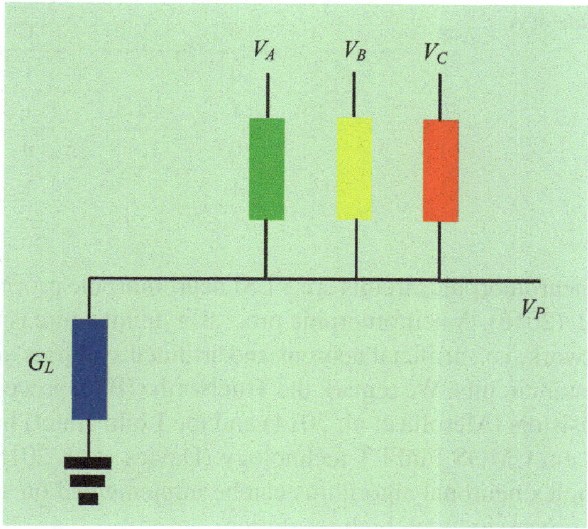

Fig. 5.33 Parallel memristor circuit for logic applications

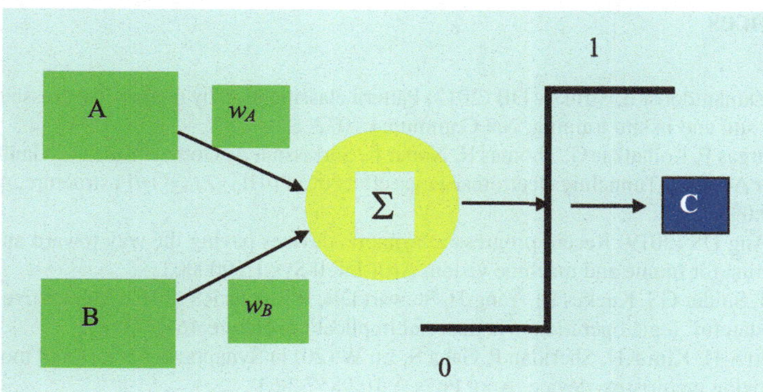

Fig. 5.34 The perceptron model of three memristors in parallel

circuit is storing the bits and performs the logical operation; this is a prerequisite for in-memory computing, discussed at the end of Chap. 3. Two memristors in parallel are able to implement the IMP gate, while three memristors in parallel implement a NAND gate. The memristors play the role of switches. Table 5.1 represents the truth table of the IMP operation, defined such that A IMP B is (NOT A) OR B.

Thus, arrays of memristors can be used to perform logical operations, as was demonstrated experimentally by Sun et al. (2018) and Borghetti et al. (2010). In the later case, an array of 1 × 17 memristors were fabricated in the form of a crossbar array, generating various Boolean logic function.

Table 5.1 Truth table of A IMP B → C′

A	*B*	*C′*
0	0	1
0	1	1
1	0	0
1	1	1

The ultimate neuromorphic circuits are VLSI neuromorphic processors, reviewed in Nawrocki et al. (2016). A neuromorphic processor architecture is that of an artificial neuronal network, i.e., artificial neurons and artificial synapses, all implemented by Si FET transistor circuits. We remark the TrueNorth (IBM) processor, containing 5 billions of transistors (Merolla et al. 2014) and the Loihi (Intel) having billion of transistors in 14 nm CMOS FinFET technology (Davies et al. 2018), with an area of 60 mm^2. Complex neuronal algorithms can be implemented on such processors, including image processing with high resolution.

References

Alibart F, Zamanidoost E, Strukov DB (2013) Pattern classification by memristive crossbar circuits using ex situ and in situ training. Nat Commun 4:2072

Ambriz-Vargas F, Kolhatkar G, Thomas R, Nouar R, Sarkissian A, Gomez-Yanez C, Gauthier MA, Ruediger A (2017) Tunneling electroresistance effect in a Pt/$Hf_{0.5}Zr_{0.5}O_2$/Pt structure. Appl Phys Lett 110:093106

Berco D, Ang DS (2019) Recent progress in synaptic devices paving the way toward an artificial cogni-retina for bionic and machine vision. Artif Intell Sys 1:1900003

Borghetti J, Snider GS, Kuekes PJ, Yang JJ, Stewart DR, Williams RS (2010) Memristive switches enable 'stateful' logic operations via material implication. Nature 464:873–876

Chang T, Jo S-H, Kim KH, Sheridan P, Gaba S, Lu W (2011) Synaptic behaviors and modeling of a metal oxide memristive device. Appl Phys A 102:857–863

Chanthbouala A, Garcia V, Cherifi RO, Bouzehouane K, Fusil S, Moya X, Xavier S, Yamada H, Deranlot C, Mathur ND, Bibes M, Barthélémy A, Grollier J (2012) A ferroelectric memristor. Nat Mater 11:860–864

Chen L, Li C, Huang T, Chen Y, Wen S, Qi J (2013) A synapse memristor model with forgetting effect. Phys Lett A 377:3260–3265

Chen Y, Gao G, Zhao J, Zhang H, Yu J, Yang X, Zhang Q, Zhang W, Xu S, Sun J, Meng Y, Sun Q (2019) Piezotronic graphene artificial sensory synapse. Adv Funct Mat 29:1900959

Chua L, Sirakoulis GC, Adamatzky A (eds) (2020) Handbook of memristor network. Springer Switzerland

Davies M, Srinivasa N, Lin T-H, Chinya G, Cao Y, Choday SH, Dimou G, Joshi P, Imam N, Jain S, Liao Y, Lin C-K, Lines A, Liu R, Mathaikutty D, McCoy S, Paul A, Tse J, Venkataramanan G, Weng Y-H, Wild A, Yang Y, Wang H (2018) Loihi: a neuromorphic many-core processor with on-chip learning. IEEE Micro 38:82–99

Garcia V, Bibes M (2014) Ferroelectric tunnel junctions for information storage and processing. Nat Commun 5:4289

Guo R, Zhou Y, Wu L, Wang Z, Lim Z, Yan X, Lin W, Wang H, Yoong HY, Chen S, Ariando Venkatesan T, Wang J, Chow GM, Gruverman A, Miao X, Zhu Y, Chen J (2018) Control of synaptic plasticity learning of ferroelectric tunnel memristor by nanoscale interface engineering. ACS Appl Mater Interfaces 10:12862–12869

Guo R, Lin W, Yan X, Venkatesan T, Chen J (2020) Ferroic tunnel junctions and their application in neuromorphic networks. Appl Phys Rev 7:011304

Hong XL, JiaJun Loy D, Dananjaya P, Tan F, Ng CM, Lew WS (2018) Oxide-based RRAM materials for neuromorphic computing. J Mater 57:8720–8746

Jo SH, Ting C, Ebong I, Bhadviya BB, Mazumder P, Lu W (2010) Nanoscale memristor device as synapse in neuromorphic systems. Nano Lett 10:1297–1301

Kendall JD, Kumar S (2020) The building blocks of a brain-inspired computer. Appl Phys Rev 7:011305

Kim M-K, Lee J-S (2019) Ferroelectric analog synaptic transistors. Nano Lett 19:2044–2050

Kim Y, Chortos A, Xu W, Liu Y, Oh JY, Son D, Kang J, Foudeh AM, Zhu C, Lee Y, Niu S, Liu J, Pfattner R, Bao Z, Lee TW (2018) A bioinspired flexible organic artificial afferent nerve. Science 360:998–1003

Kuzum D, Yu S, Wong HS (2013) Synaptic electronics: materials, devices and applications. Nanotechnology 24:382001

Mead C (1989) Analog VLSI and neural systems. Addison Wesley

Merolla PA, Arthur JV, Alvarez-Icaza R, Cassidy AS, Sawada J, Akopyan F, Jackson BL, Imam N, Guo C, Nakamura Y, Brezzo B, Vo I, Esser SK, Appuswamy R, Taba B, Amir A, Flickner MD, Risk WP, Manohar R, Modha DS (2014) A million spiking-neuron integrated circuit with a scalable communication network and interface. Science 345:668–672

Milo V, Malavena G, Monzio Compagnoni C, Ielmini D (2020) Memristive and CMOS devices for neuromorphic computing. Materials 13:166

Nawrocki RA, Voyles RM, Shaheen SE (2016) A mini review of neuromorphic arhitectures and implementations. IEEE Trans Electron Dev 63:3819–3829

Oh S, Kim T, Kwak M, Song J, Woo J, Jeon S, Yoo IK, Hwang H (2017) $HfZrO_x$-based ferroelectric synapse device with 32 levels of conductance states for neuromorphic applications. IEEE Electron Dev Lett 38:732–735

Oh S, Hwang H, Yoo IK (2019) Ferroelectric materials for neuromorphic computing. APL Mater 7:091109

Sanchez Esqueda I, Yan X, Rutherglen C, Kane A, Cain T, Marsh P, Liu Q, Galatsis K, Wang H, Zhou C (2018) Aligned carbon nanotube synaptic transistors for large-scale neuromorphic computing. ACS Nano 12:7352–7361

Sangwan VK, Hersam MC (2020) Neuromorphic nanoelectronic materials. Nat Nanotechnol 15:517–528

Shen AM, Chen CL, Kim K, Cho B, Tudor A, Chen Y (2013) Analog neuromorphic module based on carbon nanotube. ACS Nano 7:6117–6122

Sun Z, Ambrosi E, Bricalli A, Ielmini D (2018) Logic computing with stateful neural networks of resistive switches. Adv Mat 30:1802554

Vourkas I, Sirakoulis GC (2016) Memristor-based nanoelectronic computing circuits and architectures. Springer, Cham

Wang L, Lu S-R, Wen J (2017) Recent advances on neuromorphic systems using phase-change materials. Nanoscale Res Lett 12:347

Yoong HY, Wu H, Zhao J, Wang H, Guo R, Xiao J, Zhang B, Yang P, Pennycook SJ, Deng N, Yan X, Chen J (2018) Epitaxial ferroelectric $Hf_{0.5}Zr_{0.5}O_2$ thin films and their implementations in memristors for brain-inspired computing. Adv Funct Mat 28:1806037

Zheng Q, Wang Z, Gong N, Yu Z, Chen C, Cai Y, Huang Q, Jiang H, Xia Q, Huang R (2019) Artificial neural network based on doped HfO2 ferroelectric with multilevel characteristics. IEEE Trans Electron Dev Lett 40:1309–1312

Zhu X, Lu WD (2018) Optogenetics-inspired tunable synaptic functions in memristors. ACS Nano 12:1242–1249

Chapter 6
Perspectives

Abstract This last chapter describes the perspectives opened by this book.

There are many challenges regarding the continuation of Moore's law (Mitchel Waldrop 2016; Toumey 2016), which is in place since a half of century. Moore's law is not a physical law, but a technological prediction about the cramming of transistors on a chip. The law is predicting that the number of transistors on the same chip area must be doubled each 18 months, i.e., that the dimensions of the transistors must be shrinking with 50% in the same time. The most notorious adverse effect of this continuous reduction of transistor dimensions on chip, and hence of increasing their number, which reaches nowadays more than 5 billions on a single chip (Merolla et al. 2014), is neither short channels effects due to quantum mechanical phenomena, nor the limits of optical lithography, but one of the most common physical effects encountered in nature: the heat. The dissipated heat generated by billions of transistors interconnected with tens of billions of metallic interconnections can no longer be evacuated with simple cooling devices. Heat dissipation is the main cause why the central unit processor (CPU), i.e., the microprocessor speed, is limited at few gigaHertz and has not further increased since two decades. Today, each microprocessor has a couple of cores, fabricated in according to Moore's law, which are less used every generation. Thus, preserving Moore's law and increasing the number of cores has a dramatic effect. Some core areas are powered-off all the time to prevent excessive heating, the termed used being "dark silicon" (Esmaeilzadeh et al. 2013).

Another challenge regards the future of communications, the 6G at the horizon of 2030 aiming to an ever-present virtual existence (Dang et al. 2020). This goal is attained in the vision of 6G communications for the years 2030–2040 by a dramatic increase of the speed of data transmission, from 35.5 Gb/s (5G) to 100 Gb/s–1 Tb/s, meaning that the frequencies of 5G, which are 90 GHz now, will jump to the range of 1–10 THz. There are few electronic devices able to work beyond 300 GHz, i.e., in the THz range, all of them based on AIII-BV semiconductor heterostructures, which are not CMOS compatible, and rather difficult to grow and test at a massive scale.

So, the main applications of nanoelectronics, namely the devices and circuits for computers and communications, have already attained the physical limits of their

M. Dragoman and D. Dragoman, *Atomic-Scale Electronics Beyond CMOS*,
https://doi.org/10.1007/978-3-030-60563-6_6

functionalities. How can we proceed further? The answer is not simple and the entire book aims to provide an overview of various solutions to these questions.

It is well known that new materials and new device and circuits architectures have boosted the Moore's law and all applications in electronics since more than half century. A simple example proves this statement: in the twenty century 11–15 elements from the Periodic Table of Elements were used in semiconductor technology, and now they are beyond sixty (Green et al. 2017). This huge development of new materials and circuits architectures for nanoelectronic chips lead to the implementation of nanotechnologies into computation, communications, medicine, and many industries, such as automotive or aerospace industries. Nanotechnologies embody a wealth of physical concepts based on quantum mechanics and material science valid for structures having at least one dimension reaching now the atomic scale, and even the atomic level.

Therefore, the first two chapters of the book are dedicated to atomic thin materials, their physics, growth strategies and main applications. The major problem of atomic-thin materials is their growth at the wafer scale without defects. Very few 2D materials, such as graphene or MoS_2 monolayers, can be grown at the wafer scale, but defects are usually downgrading the performances of devices based on them. Electric contacts and low contact resistances are also thorny problems for any two 2D material. Therefore, a large part of Chap. 1 is dedicated to these unsolved issues, and is followed by the physical properties of 2D materials. The last part is dedicated to atom-like structures fabrication by bonds, dopants and artificial lattices, preparing the reader for Chap. 2, which is dedicated to atomic electronics—the ultimate miniaturization of any electronic device. However, there is still a long way until the moment when single dopants, atomic gap devices, such as memristors, atom transistors and graphene nanoribbons, will be currently used in electronics, due to notorious difficulties to fabricate them in reproducible manner at large scales. Nevertheless, this is the main perspective for electronics in decades to come: the atom electronics, electronics at atom scale, where in an electronic device one or few atoms only are involved.

While the first two chapters are focused mainly on 2D materials and atomic devices, the third chapter is dedicated to nanoelectronic devices based on atomically thin materials. Devices at the cutting edge of nanotechnology are presented. Ballistic transistors and other devices are very promising for ultrafast applications, but were neglected for long time since common semiconductors and semiconductor heterostructures have a very small mean-free-path at room temperature. However, the discovery of graphene has opened up a wide perspective not only on physics but also on nanotechnology: the graphene monolayer has a mean-free-path of 400 nm at room temperature, which allowing the realization of various devices, such as transistors, quantum gates, Josephson junctions, with unprecedented performances. Ferroelectric transistors and devices have now a rapid development as well, since it was discovered that very thin films of doped Hf, having a thickness in the range of 1–6 nm, are ferroelectric. This allows the fabrication at the wafer scale, in a CMOS compatible manner, of transistors with SS lower than 60 meV/dec, memories and many tunable microwave devices. Phase-change transistors, i.e., Mott transistors or tunneling transistors, are briefly presented as the future transistors for low power applications. Quantum dots and single-electron transistors (SETs) are extensively presented in connection with 2D materials, since they are key devices for quantum computing. Memristors are indispensable devices for future memories and computing, and therefore are presented in details. Devices based on 2D materials working at THz frequencies are the future devices for 6G communications and ends this extensive chapter.

The last two chapters are dealing with quantum computation and neuromorphic computation, respectively. Both are the most promising computational methods developed to continue the digital computation era, and constitute hot topics in applied research. Neither quantum computation nor neuromorphic computation are universal methods for computing, however. Quantum computation is an ultrafast computing method only in the case of some algorithms known in the literature. Since the quantum computers build these days on Josephson junctions are working at milliKelvin temperatures and consuming millions of dollars for proper functioning, they are not widespread and are used only by giant companies of the information technology era. A huge step forward will be to perform quantum computation at room temperature, and this is possible using photonic computation. However, photonic computation suffers from the lack of simple tunable devices, which are commonplace in electronics, such that the programming of such optical computers is still difficult. On the other hand, neuromorphic computation is working at room-temperature, mimicking the brain, which is based on neurons and synapses, so it is an analogue method of computation. The processors that fulfill these tasks are among the most advanced VLSI circuits, containing few billions of transistors. The neuromorphic computation is focused on neuronal algorithms, dedicated to artificial learning algorithms, vision, and image processing. New devices, such as memristors, which are acting as artificial synapses, are expected to replace, each of them, a couple of transistors performing the same function.

So, the perspectives are promising, but a lot of work is necessary for these hot topic research to become reality. Moore's law will not vanish; our only hope to go beyond this law is to "think out of the box". The devices mentioned in this book are such example. Although we cannot still replace the extremely efficient CMOS technology with something else in the near future, we will probably do so in the decades to come. As such, the book is an invitation to continue the present-day struggle for a faster and hopeful greener nanotechnological world.

References

Dang S, Amin O, Shihada B, Alouini M-S (2020) What should 6G be? Nat Electron 3:20–29

Esmaeilzadeh H, Blem E, Armant RSt, Sankaralingam K, Burger D (2013) Power challenges may end the multicore era. Commun ACM 56:93–102

Green ML, Choi CL, Hattrick-Simpers JR, Joshi AM, Takeuchi I, Barron SC, Campo E, Chiang T, Empedocles S, Gregoire JM, Kusne A, Martin J, Mehta A, Persson K, Trautt Z, Van Duren J, Zakutayev A (2017) Fulfilling the promise of the materials genome initiative with high-throughput experimental methodologies. Appl Phys Rev 4:011105

Merolla A, Arthur JV, Alvarez-Icaza R, Cassidy AS, Sawada J, Akopyan F, Jackson BL, Imam N, Guo C, Nakamura Y, Brezzo B, Vo I, Esser SK, Appuswamy R, Taba B, Amir A, Flickner MD, Risk WP, Manohar R, Modha DS (2014) A million spiking-neuron integrated circuit with a scalable communication network and interface. Science 345:668–672

Mitchel Waldrop M (2016) More than Moore. Nature 530:145–147

Toumey C (2016) Less is Moore. Nat Nanotechnol 11:2–3

Index

M. Dragoman and D. Dragoman, *Atomic-Scale Electronics Beyond CMOS*,
https://doi.org/10.1007/978-3-030-60563-6

The manufacturer's authorised representative in the EU is Springer Nature Customer Service Centre GmbH, Europaplatz 3, 69115 Heidelberg, Germany. If you have any concerns regarding our products, please contact ProductSafety@springernature.com

Printed and bound by CPI Group (UK) Ltd, Croydon, CR0 4YY
30/09/2026
02225254-0010